CRAC/Ked POT\TERY

A Compilation of Some of the Internet Publications of a "PWT/PhD." (Included are some of the early origins of "The Oscilllators-in-a-Substance Model of Existence.")

Dean LeRoy Sinclair (BA, NS, PhD)

Copyright: Dean L. Sinclair, April 20, 2014

CRACKED POTTERY

" Cracked pottery can also hold water." *Old Asaztazi Proverb*

This is a compilation of some of the writing efforts of Dean L(eRoy) Sinclair. The order is somewhat random, so you may have to pick and choose which you feel to be pertinent. At some point it may rearrange into a coherent manuscript.

Maybe I should introduce myself. Here is an auto-biographical poem published on Poetry.com a few years ago. It also was the "lead," that is, first page, poem in a Poetry.com anthology, "Forever Spoken." (Well, at least, it was page one in the copy I bought. I thought I should buy at least one of their anthologies.)

The Resident Lunatic

There is this person who thinks peculiarly,
And says weird things about chemistry,
Cosmology, biology, even relativity!
He once stated, "Where measurement ends is infinity."

He thinks objects can have relative velocity
Greater than Einstein's famous limit of "C."
He states, "An exact female clone is an improbability,
It's ruled out by XX dominance variability."

"A neutron in a nucleus may be a fantasy.
Electrons and protons are enough to there be."
He's called the clock paradox a fallacy,
And says, "Change a molecule to an atom? A possibility . . ."

On one thing most people seem to agree,
"His ideas are pure, unadulterated lunacy!"

I really do not see
Why our "resident lunatic" has to be--me!

Dean L. Sinclair

Copyright ©2009 **Dean L. Sinclair**

Pottery Piece Number One

Oscillator/Substance Model, A Theory of Everything

Summary: This article presents a view of existence as being within a substance, at its triple point, consisting of separable oscillators. As such it gives new definitions of mass, energy, force, gravitation.... Included are possible explanations of "The Big Bang," Black Holes, Neutron Stars.... In terms of this model there is also a warning of possible danger from the Hadron Collider which is due to go into operation in the Summer of 2008. .

The "Oscillator/Substance Model of Everything" is a second-generation model developed from the Motion in a Matrix Model.*The key insights of that model were that the speed of light being a limiting velocity of information transfer implied a necessary medium for interaction and that the interconvertibily of mass and energy implied that they were different aspects of the same thing. The original thought was that mass coordinated to point-centered motion and energy to motion along a line. While this seems to be in general quite close to the facts, it was realized that modifying the ideas to mass being motion confined within a surface and energy being a less specific term covering all types of motion, with Kinetic Energy being the type that would be dissipated along a line would be a more usable view.

If, instead of postulating a fixed matrix of dots as a basis for a model, it be assumed that there be a basic substance which will remain at, or return to its triple-point, the problem of whether the model is dealing with a solid, liquid or gas disappears. At the triple point the substance can act as any of the three depending upon slight variations in conditions. We may even postulate that the triple-point temperature is approximately the temperature of outer space, about three degrees Kelvin.

Postulating a substance basis for everything, allows us to define a number of things which are otherwise undefined or have "circular definitions." Mass becomes the balance between the motion content of the substance within a particular surface and the remainder of the substance. Kinetic Energy is the motion dissipated into the "Bulk" of the substance when a portion of the substance within a surface changes position. Light and other electromagnetic radiation is motion within the substance. The law of forces, "For every force there is an equal and opposite force," falls into place as a statement of the fact that if there is a pressure change within one part of the substance, there will be compensatory changes within the rest of the substance. It has been long known that pressure is a force. We simply change the statement slightly to, "All true forces are pressures."

Gravitation loses its place as a force--a status which it actually never should have had. It becomes a description of the difference in pressure of the amount of the substance between two entities and rest of the substance, including that part of the substance included within the entities. Electro-magnetism is the observed result of pressure disturbances within the substance from actions and interactions between electrons and protons,

Electrons and protons are known to be the basic units from which all of our chemical substances are derived. That is, with the possible exception of the "short-lived" particles produced in high-energy, particle-physics experiments. In our Universe, protons and electrons occur as the decay product of neutrons. Electrons and anti-electrons also appear in "Pair-production," and disappear in "Annihilation."

Putting the above pieces together with the characteristics of oscillators. we may postulate that 'Pair-production" and "Annihilation." are opposite processes. A spherical cavity oscillator can be considered to consist of counter-rotating halves, which, if subjected to enough "motion disturbance," can split into two mirror-image vortex oscillators having opposite spin orientations. If these two halves were to re-encounter on a proper axis they could recombine to the original oscillator with loss of kinetic energy. That is, postulating a basic unit of our substance as an oscillator that can be split into vortex oscillators having opposite spins, neatly accounts for "Pair-production" and "Annihilation."

As to electrons, protons and neutrons, we can account for these at the same time as we account for the "Big Bang" and the "Expanding Universe."

Neutrons fit the characteristics of a substance distorted by a shock wave such that the "front half of the unit met the back-half coming forward." This distorted unit splits into an electron and a proton. The speed of light in the substance--if light is a typical information carrier--is the average velocity in any given direction in any given instant. Therefore, it will be the average expansion velocity of the expansion phase of an oscillator, and will also be the average rotational velocity measured at the outer edge of the expansion (or contraction) phase of an oscillator. As such it will be the average velocity felt by our basic oscillators.

We may postulate that motion at above this velocity will travel as a shock wave and since "c," the speed of light, is an average, the initial velocity at the start of the expansion phase has to be greater than "c." if we consider acceleration and deceleration to be linear. we can estimate the initial shock wave from any oscillator as having a velocity of "2c" which fits fairly well with the comparative "masses" of the electron and proton. with the proton half laking the entire distortion from a "2c" shock wave slamming into our substance acting as a "solid."

What this all adds up to is that some oscillator is operating at ultra-low frequency, very high power, and we are riding in the "Creativity out of Chaos" behind that shock wave which converted may substance units into neutrons which disintegrate into electrons and protons. The electrons and protons, vortex particles of opposite spin orientations, but very different sizes and shapes, can not stably recombine but can associate in a multitude of ways, creating everything that we know in our Universe. Presumably, or the other side of the Parent Oscillator, is an "Anti-Universe, coping in its own way with the same "Creativity out of Chaos" attempt to stabilize. At some point the shock wave will expend itself, and there will be a contracting Universe-Anti-universe set which will eventually collapse through the parent-oscillator to start the process all over again.

Our discussion above leads to the conclusion that our basic oscillators possess the ability to coordinate vast amounts of energy. Taking this into account gives a possible explanation for "Black Holes" in the centers of Galaxies. In the Creativity of Chaos behind the shock wave, some units fall into coordination with an oscillator at their exact center, and begin to move with its frequencies. Considering the characteristics of oscillators, a vortex oscillator such as a proton or electron, if passed intact through a full-cavity oscillator would have its "sense" reversed. That is, an electron would become an anti-electron, a proton an anti-proton. Since, tossed out the other side of the oscillator these could unite with "untransformed" units, oscillators operating in this way would be reducing electrons and protons back to unit-oscillators. This. however, comes at a cost. Protons, formed at

distortion of smaller oscillators, would be combined into "proton-parent" oscillators, which could presumably be, in turn, distorted by shock waves into Super-neutrons, which would collapse into an anti-proton ("Super-electron") and a "Super-proton" and so on and on and on...
When a Black Hole Oscillator has done its "trash converter" work in a volume, what would appear to a human observer would be a complete blank spot in space devoid of anything which we would consider "Matter."

One other factor in our Existence that we have not mentioned is the neutrino-anti-neutrino combination. It seems logical to guess that these are a pair-product just as electrons and anti-electrons are, and that their "parent-oscillator" is a more basic unit to our substance oscillator(s). This would be in analogy to atoms and molecules in chemical substances. These could, in turn be composed of even smaller oscillators, ad infinitum.

We consider all "matter" to be made up of protons and electrons, which combine to form atoms, atoms to form molecules, molecules to form organisms.... In this model we would not consider that neutrons would exist within atoms other than in "potential" or "evanescent" state, and consider that neutron stars and quasars would perhaps model as hugely over-sized atoms. As such they would function as "isotope factories." Isotopic distributions within galaxies may be a result of differences between the "neutron" stars active within them in the past. It should be possible to find spectra of elements including possibly trans-Uranium elements or even unknown "Trans-Uraniums" in the radiation from quasars and neutrons stars. Because of the expectation of high energy electron "orbitals" closely spaced, it can be expected that these entities will have distinctive radio-frequency "signatures.

 This model, if valid, casts considerable doubt on both the validity of the assumptions behind some of the publicized experiments that are intended for the "Hadron Collider" under Switzerland and France that is due to go into operation this Summer (2008), and on its safety. If the simple definition of mass which arises in this model be valid, then the idea that something known as the Higgs Boson is responsible for mass becomes a total misconception. The idea that colliding protons with protons would duplicate the conditions before the "Big Bang" is inconsistent to this model in several ways. One is that by this model, protons would come into existence as a result of the decay of neutrons. As neutrons have a significant half-life, protons would not exist until after the start of the shock wave. As to Lead Ions, these would appear much later. However, collisions at relative velocities of close to "2c" could certainly have the potential to distort oscillators cause neutron formation and, perhaps, additional esoteric, short-lived, alternate forms of matter. Almost surely, Hydrogen would appear within the collider tube. A much more dangerous implication of this model is the idea of Black Hole formation by coordination of a circular spinning entity with an oscillator in its center. The Collider apparently has as least a couple of perfect circle dimensions which have resonance frequencies. There is a high probability that these circles will act as cavity oscillators. If one of these oscillators did come into active coordination with a central oscillator the result would be the same as is postulated for galaxies. Although on a miniature scale to a galaxy, a Black Hole interaction the size of the Collider could have results ranging from, hopefully, merely annoying "unexplained energy anomalies," to a new volcano where the Collider had been, to a coordinated vibration with the resonance frequencies of the Earth to create a new asteroid field between Venus and Mars. Although a "Black Hole Oscillator" would probably "eat" only the material within the circumference of the Collider, the conversion of that amount of matter to anti-matter would make the most powerful atomic bomb appear a mere firecracker.

The scientists working on the project apparently admit that they have no idea what will happen. I hope they remember that, in retrospect, the physicists working on the Manhattan Project realized that it was by the merest of good luck that they did not blow themselves, and the entire Chicago area off the face of the Earth.

------
* See "Motion in a Matrix...." on www.helium.com. Several other articles by this writer, which are posted on that web site also bear upon ideas related to this model.

AUTHOR'S POST-SCRIPT: After reading through the above article, the writer has realized that he has not included some things which usually occur in a modern, scientific article. There are no mathematical formulations that might not be comprehensible to everyone, nor any mention of one or more of the modern, accepted theoretical ideas such as Relativity, Quantum Mechanics, String Theory. Rather than break into the body of the above to correct these apparent flaws, there is being appended this addendum.

It has been shown in other papers that by equating the value of Planck's Constant,"h." to the dimensions of its definition, evaluating this at the speed of light, "c," one can obtain a definition of a family of dual-cavity and/or vortex oscillators defined, in set notation, by {m x r = h/c}. This set has a central, symmetric, spherical--or ring--oscillator defined by $m = r = (h/c)^{0.5}$., where "m" is a mass and "r" is the radius of a circle or a sphere. The value of m in cgs units is about 7.4×10^{-19} g. and the value of r is about
7.4×10^{-19} cm. If one goes a step farther and writes, "$m = h/c^2$" , one finds an associated minimal mass of about
7.5×10^{-58} g. which may be the mass of a "neutrino parent." Possibly the smallest mass of significance in our Universe.
When the accepted mass of the electron is inserted into the oscillator-family-set equation, the radius is found to be the Compton Wave Length used in Quantum Mechanics. Inspection of the mathematical definitions shows that the two are mathematically identical although their different derivations do not show this clearly.

Quantum Mechanics is said to fail at about 10^{-18} cm., and String Theory gets lost into its ten dimensions at about the same value. That is the value of the diameter of our average oscillator, a value through which all the oscillators may well invert. It is interesting that it is the point of failure of two of the widely used theoretical models. If our speculations are correct the failure of both is perhaps that their mathematics only accounts for half or less of "Reality."

In so far as the writer knows, no-one has correlated the Space-Time approach with "Q.M." Each seems to have been in its own niche. The mathematics of both, in so far as they go, would seem to be somewhat compatible with this model. Space-Time modeling seems to work quite well at the macro level despite the fact that Relativity probably only would strictly
apply to communication theory, and concepts of what occupies space and what time actually is, appear to be in error. String Theory seems to depend, ultimately, on the true, but-rather-useless-in-the-thinking-of-most-people fact that the route traced by a dot moving in random fashion will follow a path which one can consider as having always being there. The 10 dimensions may arise because one can use nine dimensions of space, and one of motion. to define the actions of a dot--if one wishes to do so.

If the foregoing seems not too well explained and a bit abstruse, then, perhaps, the writer has reached the goal of writing a "modern, scientific paper." Most modern, scientific papers seem to be written in such a way that only the author and the closest associates have any idea what is going on. If they do.

On a more serious note, it can be observed that if this model be valid, it could have been put together close to a century ago, had the ideas advanced by Planck been followed up rather than those of Einstein. This writer did not realize this at the start of the path that led to this write-up. However, the initial impetus was the realization that there was an "Overlooked Obvious" in Einstein's Special Relativity, the fact that the transform equations could be generalized to fit into communication theory if one realized that "c," the speed of light was simply the limiting velocity of an information carrier wave. The theory was valid for situations wherein communication was possible or necessary; but, possibly of no pertinence in other situations.

Fragment number 2. **Old Data/New model**

This is aslightly different view of essentially the same material. Included are comments by another editor, a nuclear physicist, a decade younger than this writer, who teaches in India.

OLD DATA/NEW MODEL http://docs.google.com/Doc?docid=dcb2474d_233q69gfghq   Dean L. Sinclair

Reversing the century old interpretation of the Michelson-Morley Experiment and renaming Planck's Constant from a "Constant of Action" to the equivalent, "Constant of Angular Momentum, " leads to a new model of existence, a simple model which may be comprehensive.

The Michelson-Morley Experiment of 1890 showed that the Speed of Light. "c." was a Constant of Nature (as nearly as could be determined) . The interpretation that was given to this--Albert Einstein seems to have been a major proponent of the view--was that this fact proved that the "Aether," the all-pervasive something thought to be everywhere, does not exist. Light waves simply pass through empty space at the fastest possible speed at which anything can move. This view seems to be accepted by the scientific community to this day.

An alternative, almost totally reversed, interpretation goes something like this: The Michelson-Morley Experiment defined the maximum speed at which information is carried in the "Aether," whatever it may be. It, also, showed that the "Aether" can act as if it were a solid, carrying the transverse wave disturbances which we understand as "light."

I like the generalization. Also, the "solid" is a good point, since liquids cannot support transverse waves!

The maximum speed of information transfer from a point can be shown to be the average speed in any direction of the motion of the information carriers, whether these carriers be Pony Express Riders or Electromagnetic Waves. Considering this, "c," the Speed of Light, would be the average velocity, measured in any direction, of the motions pertinent to the carrying of information.

Max = average might cause/need some discussion.

If one sends information, no matter how, it is passed by "packets" which pass from one carrier to another. Using the Pony Express analogy, if packets of mail were sent out in every direction from a center, the maximum speed of transfer would be the average speed that the ponies and riders could maintain in any given direction. The same idea holds for any type of information transfer. ds

At about the same time as the Michelson-Morley Experiment, Max Planck was working out his famous relationship between the frequency of light waves and "Energy." The result is well known. It is the equation, "Energy equals Planck's Constant times the frequency. That is "E=hu." E can be expressed in "Ergs" with, "h," the constant, in Erg-seconds, and "u, Nu." in cycles per second. Erg-seconds is called "Action" and Planck called his constant the "Constant of Action." Although action has the same dimensions as "Angular Momentum," this fact was ignored. Had Planck called his constant the "Angular Momentum Constant--" which would have emphasized a rotatory aspect to Energy and Wave Motion--scientific theory might well have developed in a totally different direction.

I agree!

If we take Planck's Constant as being an Angular Momentum Constant of Nature and equate it to the definition of angular momentum, "Mass times radius of the rotor times the velocity of the rotor measured at the outer edge ("tangentially"). we obtain the equation, "h=mrv." Since this constant applies for the movement of Information/Energy at the speed of light, "c," it makes sense to evaluate this at the velocity, "c." Doing this produces the equation, "h=mrc," which can be rearranged to "mv=h/c." This valuable little equation defines "h/c" as a "Torque Constant of Nature." What more it discloses will be explained shortly, as applying a simple mathematical principle to the above leads to a new Model of Existence.

Should be "h=mrc," instead of "h=mvc."
Thanks for catching the "typo!" I corrected it. ds

 Any equation in two unknowns of the form, "xy=K," two unknowns multiplied together equals a constant," can be rewritten in the form, AxBy=K=BxAy. That is the values of the "coefficients" of the units of the variables can be interchanged to give a valid statement. In this form, this little mathematical equation has much use in physics, It is the Law of Levers, the Law of Balance, the Law of Forces, and can be used as a defining statement for the Limits of an Oscillator,,,, Noting this use in Oscillator Definition, we see that the equation, "mr=h/c," is an equation of the type discussed above; and, therefore, can be considered to be the defining equation for a family set of constant torque oscillators, {mr=h/c}, with a torque of "h/c" and an average "size" at the value wherein the 'Absolute Values" of mass and radius are equal, that is when, "m = r= (h/c)^0.5." (4.7 x 10^-19 is the value, to two significant figures, found when (h/c)^0.5 is evaluated in "cgs" units.)

 Torque = F r = m a r where a is an acceleration. Torque may not be the term you should be using!
   What was done here was mixing what were being considered essentially identical definitions. In the definition of angular momentum using radian measurement, mass and radius are combined into "torque," if I remember rightly.ds

By re-interpreting the Michelson-Morley Results and taking an alternative view of Planck's Constant, we have come to an alternative model of existence as within a something, "Aether," consisting of rotating units, which are--or are organized into--oscillators of the family set, {mr=h/c}. We have found an average value for these oscillators and, have found a mathematically defined "dimension," previously unreported, which would apply to all of these oscillators, this would be the "dimension" of masses greater than 4.7 x 10^-19 g. and radii less than 4.7 x 10^-19 cm.

One major question remains about this as a model. How can the "Aether" act as a solid? A look at chemical theory answers this question. If the "Aether" be considered to be a substance at its triple point--where it can be solid, liquid or gas--this problem disappears. We have a model, therefore, which may be called and "Oscillator/Substance Model."

You have identified an good point. A normal gas or liquid cannot support transverse waves. But space is not a solid. Your hypothesis of it being at a triple point is ingenious, but not the only option. I see 2 others right off.
   1. There is a 4th phase (twisters?).
   2.

The "triple point" analogy is as close as I could come to the idea of a "something" which would tend to equilibrate about an "average" motion content, wherein motion disturbance at the average would be transmitted in a manner as if the substance were a solid. We also need a something wherein collected motion disturbances, such as the one writing this, can move about....The twisters is an interesting comment since the "Substance" seems to be organized into "oscillators" certain of which, the "charged particles" certainly turn and twist.....ds

2. The aether is a quantized "gas." (or medium)

   No argument here. A 'quantized medium" certainly seems as if it would fit....ds

 These may actually reduce to the same thing. Quantum mechanics might find the 2nd option very appealing to study.

You might find my article on SciScoop, "Quantization, a 3-DMerrygoround?" interesting. It is also reprinted on the web site noted below. ds

There is an open membership web site,
http://groups.google.com/group/oscillatorsubstance-theory set up to explore, elaborate, and/or refute the implications of this model.
 (You are all, of course, invited to participate in that site to any extent that you might wish.)ds

Follow-ups on this model give some potentially quite useful and informative views on a number of topics. Here are some examples:

     Law of Forces (Equal and Opposite)--Balanced pressures. Also, mathematically, as above, AxBy=K=AyBx.

     Force of Gravity--Not a true force, an observational fact explained by differential pressures, even if it can be, and is, handled mathematically as an "attraction."

     Mass--Balanced pressures/tensions at a surface.

     Negative and Positive Charges--Result of counter-clockwise and clockwise spins of "stable" vortex oscillators/oscillations in the "Substance." usually the electron or proton.

Cannot be a rotation in 3-space, otherwise the charge will change with orientation. I believe that it is a rotation about the time axis and is therefore the same independent of orientation wrt the space axes.

We'll have to thrash this one out. I think that we are saying essentially the same thing. The rotation, inversion apparently is a continuous cyclic process which will force the unit to always be moving in "3-D space, as a unit." The rotation being once per inversion accounts for the 720 degree inversion noted in a fairly recent paper. ds

Electron--Counter-clockwise-spinning, inverting-vortex oscillator/oscillation having oscillation limits of 9.11×10^{-28} g. at 2.43×10^{-10} cm. and 2.43×10^{-10} g/ at 9.11×10^{-28} cm. (Note: Rotation sense is a guess.)

Proton--Clockwise-spinning, inverting-vortex oscillator/oscillation having oscillation limits of 1.67×10^{-24} g. at 1.32×10^{-13} cm. and 1.32×10^{-13} g. at 1.67×10^{-24} cm. (Note: The electron is "larger and lighter" than the proton in the measurements that are reported in the scientific literature. It is "smaller and heavier" than the proton in the "hidden half of an oscillator" smaller than 4.7×10^{-19} cm. in radius.)

I think that you have deviated from the best path at this point. I believe that you should consider the positron as the clockwise rotator (and the **co-first** fundamental particle) and protons and neutrons as composite particles.

The positron would be a clockwise rotator. In a sense all of these units would be composites. It can be shown mathematically, that if a positron were to be slowed to about 1/42 of its translational motion with all of that translational motion converted to "mass" it would be a proton. I don't understand your notation of "co-first."ds

Neutron--A unit which is postulated to be formed from a "shock-wave-squashing-process" of the "Zerotron." It may be considered as a dual oscillator composed of a tightly coordinated electron-proton pair. The much slower vibration of the electron would tend to pass both outside and inside the proton in a tight coordination in the neutron. If this be a valid model.

Neutrino--One of several postulated particles associated with charged particles which are considered to be necessary for the conservation of momentum. They may actually be a description of a part of the motion of the particular oscillator with which they are associated. (The internal motions of a charged particle can probably by mathematically described as if they were composed of two coordinated oscillators. The "smaller oscillator" with the the shorter vibrational range and a rotation such as to be in the direction in which the "charged particle" will continue to move" may actually account for the "neutrino," which is said to have "rotation in the direction of motion.")

Electron-anti-electron-annihilation--Combination of these two vortex oscillators of opposite rotation but otherwise identical characteristics into a spherical, pulsating oscillator, having the same mass and size limits as either of the "halves." (This was dubbed , "zerotron." See article, "Negatron plus Positron Equals "Zerotron?" posted on SciScoop . Note: This article is also posted on the Google site.)

[This "Zerotron" might be considered as possibly the most likely "fundamental unit," at least in our particular universe.ds]

Pair-production--Reversal of above process....

The "Big Bang--" The instant of inversion--or splitting--of an oscillator creating a Universe/Anti-Universe Set. Neutrons being created on "our" side, which collapse to electrons and protons....

"Zerotron--" A hypothetical spherical oscillator, probably ubiquitous to the "Substance," which is splittable to electron-anti-electron pair and deformable to the neutron and/or anti-neutron....The sense of deformation would depend on the twist of the shock wave.

The above are some of the definitions/explanations which develop easily from this model, most of which do not seem to arise from any other model. It is hoped that despite its proletarian birth from re-interpreting old data from a different view, people in the scientific community will give it a chance to prove its worth--or lack there of--without simply dismissing it "out of hand."

A very optimistic view is that Einsteinian Space-Time theorists will realize that they are actually talking about a type of Substance Field* and reexamine their calculations. It will not be surprising if they fit in very well. The Standard Model devotees might try checking to see if their various postulates might not be simplified, reinterpreted or discarded in light of this model. Quantum Mechanics could conceivably be extended to other than positive numbers to account for the "hidden half of existence smaller than a diameter of about 9.4×10^{-19} cm." Chemists and physiclsts might consider the effects of thinking about protons and electrons as oscillating units and positive and negative charges as being expressions of the different effects of reversed rotations in a medium.

This rather naive little model just might possibly be the start of a break through to unify the fields of physical science. If enough scientists were open-minded enough to consider how it might fit in....

The above is a bare bones summary of a developing model. Various aspects have been further explored in several articles on SciScoop, most of which have been also posted as "pages" on the Oscillator/Substance Theory Google Site. There are pages there, also, which have not been published elsewhere, including one on "The Equivalence of Electrons"

which have not been posted elsewhere. The earliest beginning Internet article was published in April of 2007 on Helium.com as "Motion in a Matrix...."

This seems to be the first model to delve seriously into the possibility of structure and form of electrons and protons and to give fairly easy definitions for mass, energy, gravitation....It, also, is the first notation of the "hidden half below 4.7×10^{-19} cm.," although this has been hinted at by statements of this sort: "Quantum Mechanics fails at below 10^{-18} cm." and "The strings of the string theorists disappear into a hole at 10^{-18} cm. "
------
* See paper, "Why Einstein was right" reprinted on the Google Group referenced above. ds

Piece number 3 Further amplification.

Elements, Periodic Chart and O/S

If the Oscillator/Substance concept of what are electrons and protons be valid, and electrons and protons are actually spinning vortices, with the size of the protons and their interactions accounting for the "nuclei" of the nuclear atom, then it is necessary to do a complete re-examination of the concepts of atoms, elements and the basis of the Periodic Chart of the Elements. A new reason/rationale need be developed for the apparent presence of "isotopes" which are conventionally explained by the presence of neutrons in atomic nuclei. If there be not neutrons, per se, in the nuclei, then what are the electron-proton interactions which explain a situation which will make it appear that there are stable neutron type associations?

Conventional wisdom says that there be pairing of electrons such that their "spins" will cancel. O/S would agree that a partial cancellation of the vortex motions would take place if two vortices were "paired, upside down to one another." The conventional literature, however, does not continue to apply the same idea to protons, nor does it seem to realize that an up-down-up-down chain or circle can be extended. It should be useful to consider the first few known proton-electron associations.

The simplest units which may be considered as proton-electron associations are the neutron and the Hydrogen atom, H1. The neutron may possibly be, or, at least can be though of as, a tightly coupled dual oscillator of an electron and proton, or "proto-electron/proto-proton," which when it "falls out of sync" collapses into an electron and a proton. These two can reunite into another unit which can synchronize in unlimited--or almost unlimited--spatial dimension, hence, cannot be "knocked out of sync" in the manner of its more restricted isomer, the neutron.

Perhaps the next easiest set of units to consider is what one might call "Iso-set--2,2," the set of units made up of two electrons and two protons. This would have also two members, the Deuterium Atom, "D," or, "H2," and the Hydrogen molecule, H:H. The Hydrogen molecule is known to have two forms known as "ortho-Hydrogen" and "para-Hydrogen" wherein the Hydrogen nuclei, ie, protons are "matched spin" or "paired-spins." that is the protons are moving in exactly the same orientation in one case and are "upside down to each other" in the more stable orientation. The Hydrogen molecule would be expected to be an ovoid. It is probable that the Deuterium atom could be considered as a condensed version of the "para-Hydrogen" molecule, containing much less vibrational energy. The close-coupling of the two protons in the nucleus and their probable containment, at any given instant within an electron, giving the illusion of a "nucleus"L consisting of a proton and a neutron. It can be considered that the vibrational motions of the electron and the protons would be such that for very short periods, the proton/electron interactions would be such as to be identical to those of a neutron, however, the neutron as such would not exist....

The three electron, three proton set, "Iso-3. 3-set." has two well-known, atomic members, Tritium, "H3," and Helium 3, "He3." Tritium is "radioactive," spinning off an electron to form a cation which when it regains an electron becomes the isomeric He3. Tritium does not have a magnetic moment; therefore, it is not a "spinning neutral unit." This implies an internal symmetry, if the nucleus be composed of an inverting tetrahedral array, that is, three units inverting as if they were at corners of a tetrahedron, then this unit would on the average have no polarization, do "dipole" even though the internal units have inherent dipoles. It appears from the chemistry of this unit that, at any given instant the central array, the "atomic nucleus" is encased within a set of two coupled electrons with a third electron more loosely coupled.

The "stable" He3 unit is a totally different configuration. With a definite magnetic moment in the same sense as that of the neutron, but larger, it is a spinning "neutral" atom, with a definite dipole dominated by the spinning protons which would appear to be in a trigonal array corresponding on the atomic scale to the "resonance-stabilized" trigonals that are known in molecular chemistry. These three appear at any given instant to be tightly encased in the vibrational pattern of one electron and more loosely in the vibrations of two others....This giving rise to the conventional idea of one neutron in the nucleus. Evanescent, transient states which would be identical to states of neutrons, would exist in both the Tritium and Helium3 units, as in all atoms, giving rise to the illusion of stable neutrons in nuclei.

In the "Iso-4,4-set," it interesting to compare the molecular unit, Deuterium molecule, D:D, and its atomic isomer, the He4 atom. This, however, has been covered in some detail in another short paper, "O/S Theory and the Deuterium to Helium4 Transform." Quoting from that paper:
[Deuterium Atom nucleus consists of two, "coupled-up-down" protons (in the most stable state) and the Deuterium Molecule would be two sets in what could be called a "stretched tetrahedral array" or possibly a "stretched square planar array," or an arrangement combining these two ideas. This set of "nuclei" would be surrounded, literally within and without, by an array of 4 electrons.
 The overall result being an ovoid with two distinct centers of motion.

The He4 nucleus would be composed of the same eight basic units; however, in this case the ovoid of two distinct centers of motion is compressed into a spheroid having but one center. It is possible that this spheroid would have a tetrahedral configuration of protons again surrounded within and without by a tetrahedral arrangement of electrons .

The Alpha particle would not be a He4 nucleus, but a square-planar (actually, circular) array of 4 coupled protons surrounded again, "within and without" by two electrons. The coupling of the protons giving the unit a very strong, clockwise? spin.

These "pictures" make sense if you look at the oscillator limits of the proton and the electron and the reversed spin/inversion senses which give rise to the "positive" and "negative" charges.

 (The electron and proton by O/S logic have the same "average" mass and radius, but with very different oscillation limits, with the electron limits approximately "ten to the third" times those of the proton hence, with both having the same rotational/inversional velocity of "c' the proton does many inversion/rotations for each of those of the electron. The electron is. therefore, "heavier and lighter" and "larger and smaller" than the proton. The reader is also referred to the "page" http://groups.google.com/group/oscillatorsubstance-theory/web/the-electron-and-proton-as-oscillators)]

It can be seen that the analysis of the entire periodic chart and corresponding molecular isomers would be an impossible task. It appears that nature is such that certain patterns arise of electron-proton interactions which lead to the Periodic Chart pattern with the use of "neutrons in the nucleus" as a convenient bookkeeping tool. This, however, has to be remembered, if one feels the O/S insights to be valid, is illusional and can obscure other patterns of value. some of which can be found in use of the Iso-set, Iso-A, concepts covered, to some extent, in another article. "Iso-sets, a Key to Radioactivity." which can be found on SCiScoop, and. also, as a page on the Google Group, Oscillator/Substance Theory.

Unit 4. A little back-tracking. This article was metioned before.

The Electron and Proton as Oscillators

If it be taken that the relationship between energy and electromagnetic radiation be a fundamental relationship of our universe, then examination of that relationship should furnish clues as to the nature of our universe.

If Planck's Constant, the constant which relates energy to electromagnetic radiation be equated to its definition as an angular momentum, one obtains the equation, m x r x v = h. Evaluating this at "c," the speed of light, we obtain, mrc=h which can be rearranged to mr=h/c.

As any equation of the form xy=K can be taken to describe an oscillator, by writing it in the form, xy=K=yx, to emphasize the interchangeability of the values of the two variables, we can see that the equation, mr=h/c, can be taken to define a family of oscillators of constant torque, h/c.

When the electron is checked to see if it fits into this family, it is found to fit with one limit set with the "rest mass" as the mass, "m", and the "Compton wavelength" as the radius,"r."   Noted in our Universe as "Rest Mass" and "Compton Wave Length" are, (Six significant figures)
For an oscillator the absolute values can be switched to determine what the other limit is. For this particular oscillator, the other oscillatory limit would be absolute value of Compton Wavelength as mass and the absolute value of the "rest mass" as the value of the radius, "r."

 The most consistent set of values seems to be cgs, so the following are what is obtained as the vibratory limits of the electron:
9.10953 x 10^-28 g. for the mass, correlated to 2.42631 x 10^-10 cm.
The other oscillatory limit would be 2.43631 x 10^-10 g. correlated to   9.10953 x 10^-28 cm.

We can analyze the proton in the same way to obtain the values:
 Our observations--minimal mass, maximal radius--1.67264 x 10^-24 g. and 1.321401 x10^-13 cm. ; balanced by
 the other limit of maximal mass, minimal radius-- 1.321401 x 10^-13 g. and 1.67264 x 10^-24 cm.
 It is to be noted that the electron is both heavier and lighter than the proton and larger and smaller, depending upon where the observation be taken. At the average values they are the same, as are all oscillators of this family. The logic is, as folows:

What we have discovered, if the mathematics above accurately reflect the "real world," is that what we are observing is a situation of minimal mass and maximal size of the electron which is "balanced" in what we might call something like an alternate reality, by a maximal mass and minimal size. We have also found an average size for all oscillators of this family of (h/c)^0.5. This is approximately 4.7 x 10^-19 cm. radius and 4.7 x 10^-19 grams.

As there is recent literature observation that the electron, and, presumably also, the proton, turn 720 degrees to return to their original states, it seems sensible to consider these entities as "vortexes" which invert once per rotation, hence 720 degrees to return to the original orientation.

It can be seen that the application of the mathematical concept that any set of unknowns, xyz...n = K = n...zyx can be said to define an oscillator situation in as many dimensions as one wishes to use, can have interesting implications. In this case the implication is that we possibly exist in a universe of a substance made up of oscillators of the constant torque family, h/c.

An Internet, group has been set up to explore, amplify, elucidate, or refute the implications of the above ideas. Its "URL" is http://groups.google.com/group/oscillatorsubstance-theory.

. **Five A write-up for a practical problem...**

O/S Theory and the Deuterium to Helium Transform Dean L. Sinclair
http://docs.google.com/Doc?id=dcb2474d_2499558sw66&hl=en

The Oscillator/Substance Theory which has been rather accidentally developing over the past few years, but is essentially unknown, seems to have some interesting application to the DD-->He4 question.

In O/S Modelling there is no need for neutrons to exist in most atomic nuclei, ie, any atomic nuclei having lifetimes in excess of the known life-time of the neutron. The size/shape of nuclei should follow naturally from the size, shape, and motion of the protons and electrons which make up the given unit. In O/S, electrons and protons can actually be defined by size, shape and their projected type of motion.

By this view, the Deuterium Atom nucleus consists of two, "coupled-up-down" protons (in the most stable state) and the Deuterium Molecule would be two sets in what could be called a "stretched tetrahedral array" or possibly a "stretched square planar array," or an arrangement combining these two ideas. This set of "nuclei" would be surrounded, literally within and without, by an array of 4 electrons. The overall result being an ovoid with two distinct centers of motion.

The He4 nucleus would be composed of the same eight basic units; however, in this case the ovoid of two distinct centers of motion is compressed into a spheroid having but one center. It is possible that this spheroid would have a tetrahedral configuration of protons again surrounded within and without by a tetrahedral arrangement of electrons .

The Alpha particle would not be a He4 nucleus, but a square-planar (actually, circular) array of 4 coupled protons surrounded again, "within and without" by two electrons. The coupling of the protons giving the unit a very strong, clockwise? spin.

These "pictures" make sense if you look at the oscillator limits of the proton and the electron and the reversed spin/inversion senses which give rise to the "positive" and "negative" charges.

[The electron and proton by O/S logic have the same "average" mass and radius, but with very different oscillation limits, with the electron limits approximately "ten to the third" times those of the proton hence, with both having the same rotational/inversional velocity of "c' the proton does many inversion/rotations for each of those of the electron. The electron is. therefore, "heavier and lighter" and "larger and smaller" than the proton. The reader is also referred to the "page" http://groups.google.com/group/oscillatorsubstance-theory/web/the-electron-and-proton-as-oscillators]

Shifting to a somewhat different approach, let us consider some interactions among the above units. In a skeletonized, inorganic chemistry approach, we can write the following possible set of transforms for the electrolysis reactions which seem to produce the Deuterium to Helium 4 transform.

1. Ionization: DOD --> D+ + DO-

2. Electrolysis, cathode: D+ + e- -->D.

3. Combination: D. + D. --> D:D

4. Oxidation: D:D - e- --> D.D+

5. Coupling: D.D+ + D.D+ --> D.D:D.D++

6. Disproportionation/Splitting: D.D:D.D++ --> D:D + D:D++ (Alpha particle ?)

7. Acid-Base Reaction with unit regeneration: D:D++ (Alpha) + D:D --> He4 + Another Alpha

This sequence may be too long and is, of course, grossly simplified; however, it can be seen that every reaction noted except for the initial electrolysis reaction is expected to be exothermic, and once an Alpha particle type intermediate was generated the further reactions would be expected to be self-sustaining,,..

All of this suggests that there is probably a necessary lading of the electrodes to produce a situation of Deuterium molecules in contact with one another, and also to produce enough polarization of the electrodes to force sufficient back emf to produce a few initiating Deuterium molecular cations.

If, as the above suggests, there is an Alpha particle intermediate produced, then it is possible that an Alpha emitter in contact with molecular Deuterium in a condensed state, either on a catalytic surface or as a liquid at very low temperatures could initiate this type of transform. The liquid at low temperatures idea, could, of course, result in a rather sudden reversion of the Deuterium to a gaseous state, with predictable results to the apparatus used.....

Much of this has previously been covered in another short note,
http://groups.google.com/group/oscillatorsubstance-theory/web/deuterium-molecular-cation
.

Sixth. A new complication appears,

Reduction to One Dimension--Applied to the Planck's Equation.

Dr. Brian Josephson mentioned once a distinguished teacher who would announce at the beginning of a discussion that to simplify the discussion he would set Pi and "i." equal to "One." This reminded me that one could reduce any "multi-dimensional" problem of the form. xyz...n=K=...zyx to an equation of the form, x=K by dividing out the other dimensions, essentially equating everything except the one dimension being considered to "One."

Applying this to Planck's relationship, Energy equals Planck's constant times frequency, E=hu we can divide out all of the dimensions associated with "h" leaving "h" as a pure number, if we do this, what we have left behind on either side of the equation is 1/unit of time = h cycles/unit of time, now if we even remove the time unit, we find that 1= h cycles or 1/h is a constant describing a basic cycle.

We can reinsert any time reference which we please and a value of "h" which is consistent with the system of measurements that we would wish to use were we to go back to the dimensions which we divided out. Going to second and the cgs system, we can evaluate this "cycle" constant as $1/(6.63 \times 10^{-27})$ cycles/sec. If this cycle is riding on a carrier wave at "c," centimeters/sec. then we have 1/h cycles in "c" centimeters for a wave length of hc cm. A basic wavelength associated with this equation would be (hc)cm.

Now since this wavelength could be considered as the circumference of a circle, a radius "r" can be calculate as (hc/2Pi) . this would be recognized by physicists as "c times 'h-Bar,'" where "h-Bar" is Dirac's Constant , h/2Pi.

Another analysis, by Equating Planck's Constant to its dimensions as an angular momentum, gives the equation, mrv=h. Evaluating this at "c." gives mrc=h, which rearranges to mr=h/c. If we insert hc/2Pi in this equation for "r" and rearrange, we obtain, "m=2Pi/c^2.

Going farther, since, mr=h/c=rm is the equation for the limits of an oscillator, we can say that we have defined one oscillator limit at m=2Pi/c^2 and r=hc/2Pi, the "Reciprocal limits" by the balance law or the law of levers would be m=hc/2Pi and r=2Pi/c^2. That is the dimension titles, will switch absolute values.

This suggests that the transfer of information in 'black body radiation" according to the Planck's equation, indeed of all electromagnetic radiation is associated with an oscillator having the above limits. that is essentially operating between the limits of 3×10^{-17} cm. and 7×10^{-21} g. and 7×10^{-21} cm. and 3×10^{-17} g.

This can be compared to the "average" value for all of the oscillators of this family calculated from m=r=(h/c)^0.5, of about 4.7×10^{-19} cm and grams.

These can be compared to the "Our Universe" values for the proton of 1.32×10^{-13} cm.

and   1.64 x 10^-27 g. In "Our Universe," measurements, could the unit that seems to appear above be observed, would make it seem to have about 4.3 million times the mass of the proton....

Seven.   An essential idea that does not come from Planck's Constant and the Speed of **Light.**
Negatron Plus Positron Equals Zerotron?

e^- + e^+ = e^0 ?

A little manipulation with the idea of "electron-anti-electron annihilation" leads to the consideration of there being a neutral combination particle.

The electron and anti-electron (negatron and positron) are known to combine and disappear. This gives off energy as "annular radiation" and is known as mutual "annihilation." However, let us take a look at the situation.

Let us assume that an electron spinning in one direction along a given axis at the speed of light, meets an anti-electron travelling at the same velocity with the opposite spin orientation. As the velocity of each, in opposite vectors is "c" the speed of light, the Kinetic Energy of each is
(mc^2)/2 and the Energy expected to be dissipated in the "head on collision" is mc^2. However--believing Einstein--we see that we still have at least one "mc^2" worth of energy unaccounted for as each of the two participants in the "collision" had this much Energy ascribed to them.

This indicates that there is enough Energy remaining for a particle of the same mass as either of the originals to have survived the collision. A logical thought is that, rather than a "mutual destruction," the negatron and positron simply did a "Yin-Yang" combination to a neutral unit dumping excess rotational motion as "annular' (ring-form) radiation into the plane in which they met.

There is another known process, called "pair-formation" in which radiation above a certain threshold, the same Energy as the annular radiation. will, under certain conditions, lead to the appearance of an electron and a proton. It is reasonable that a "parent entity" of the "positron and negatron," which we may call a "zerotron," "e^0," could be an explanation for both phenomena.

A neutral entity as a "parent to a positive-negative pair" is, of course, known. The neutron is the "neutral parent" to the electron and proton. It is not too far-fetched to suggest the following: e^0 + Energy --> n^0 --> e^- + p^+ + Energy. That is, that under some condition the unit which we are postulating as the predecessor of the electron and anti-electron may be converted to a neutron which then splits to the electron and proton.

Along the same lines of reasoning we may note that the Muon comes in three forms, positive, negative and neutral. Has anyone checked to seen if the positive and negative forms combine to form the neutral one? Probably that would be very hard to check.

One may also suggest that there possibly be a neutral predecessor of the neutrino and anti-neutrino, a "zeroino."

7A This same article was published on SciScoop. This is the article as posted on SCiSCOOP

Positron and Negatron Equals a "Zerotron?"

Physics **Wednesday**, December 10, 2008 *by* *deanlsinclair*

$e^- + e^+ = e^0$?
A little manipulation with the idea of "electron-anti-electron annihilation" leads to the consideration of there being a neutral combination particle.
The electron and anti-electron (negatron and positron) are known to combine and disappear. This gives off energy as "annular radiation" and is known as mutual "annihilation." However, let us take a look at the situation.
Let us assume that an electron spinning in one direction along a given axis at the speed of light, meets an anti-electron travelling at the same velocity with the opposite spin orientation. As the velocity of each, in opposite vectors is "c" the speed of light, the Kinetic Energy of each is
$(mc^2)/2$ and the Energy expected to be dissipated in the "head on collision" is mc^2. However–believing Einstein–we see that we still have at least one "mc^2" worth of energy unaccounted for as each of the two participants in the "collision" had this much Energy ascribed to them.
This indicates that there is enough Energy remaining for a particle of the same mass as either of the originals to have survived the collision. A logical thought is that, rather than a "mutual destruction," the negatron and positron simply did a "Yin-Yang" combination to a neutral unit dumping excess rotational motion as "annular' (ring-form) radiation into the plane in which they met.
There is another known process, called "pair-formation" in which radiation above a certain threshold, the same Energy as the annular radiation. will, under certain conditions, lead to the appearance of an electron and a proton. It is reasonable that a "parent entity" of the "positron and negatron," which we may call a "zerotron," "e^0," could be an explanation for both phenomena.
A neutral entity as a "parent to a positive-negative pair" is, of course, known. The neutron is the "neutral parent" to the electron and proton. It is not too far-fetched to suggest the following: $e^0 + Energy \rightarrow n^0 \rightarrow e^- + p^+ + Energy$. That is, that under some condition the unit which we are postulating as the predecessor of the electron and anti-electron may be converted to a neutron which then splits to the electron and proton.
Along the same lines of reasoning we may note that the Muon comes in three forms, positive, negative and neutral. Has anyone checked to seen if the positive and negative forms combine to form the neutral one? Probably that would be very hard to check.
One may also suggest that there possibly be a neutral predecessor of the neutrino and anti-neutrino, a "zeroino."

I tried to copy the following story from SciScoop, also, but for some reason it

would not copy, so I had to go to my own, no-where-near-as-pretty files.

Why Einstein was right.

Why Einstein was right, when he was, in his "Relativity" theorizing, was often because hidden facets of mathematics and of the definitions used in physics allowed the ideas to be usable, even if his reasonings were fallacious. In other words, often, he was lucky.

Einstein's Special Relativity, which title for it I am told he disliked, accurately describes how information is changed as the relative velocity of a transmitter/receiver pair approaches the velocity of the carrier wave. When the carrier wave velocity is taken to be the limiting relative velocity of any two independently moving objects, the Special Relativity view leads to nonsensical conclusions.

"SR" accurately predicts that "Mass" will increase when a moving object reaches the "speed of light." However, the prediction is that the mass will go to "Infinity." There are some problems with that. Infinity." in practice, simply means that our measurement device, or our logic, fails at this point. There are two other factors, of which Einstein seems to have been unaware, which come into play here to allow his model to accurately predict a change in the situation at the speed of light.

First, even mathematics does not allow empty space. There is always an implied "dot field" or "dot matrix," so it can be expected that what ever is considered to be moving would be moving within "something."

The other factor is also mathematical. When the process of "Integration " is carried out on the momentum equation, mass times velocity equals momentum, ($m \times v = P$), this process can be carried out with either the mass or the velocity considered as the variable. If velocity varies, we get the usual kinetic energy equation, $KE=(mv^2)/2$. However, if mass be considered as the variable, with velocity constant, we obtain another energy equation, $(vm^2)/2$. This latter does not appear in the literature and has apparently been ignored. These two equations may be interpreted as indicating that "velocity" can change to a limit, if it hits a limit, and the "accelerating situation" continues, then, "mass" will change.

Neither the "dot-matrix" aspect of mathematics, nor the alternative energy formulation, seems to appear in any of Einstein"s work, nor anywhere else in the readily available literature. If both the dot-matrix and the alternative energy expression that appear in the mathematics are reflected in reality, we see that any moving entity within the matrix will, itself, be a part of the matrix and the mass will be a measure of the balance of the interior of the moving entity and the remainder of the matrix. As long as the translational velocity of the moving object is small with respect to the average speed of motion in the matrix, there will be little effect of the "second" energy equation. Velocity will change and the amount of disturbance dissipated by the matrix, i.e., the "Energy" will be measured quite accurately by the "Kinetic Energy" formula. There is a change in the situation when the translational velocity of the moving object starts to equal or to exceed the average of the matrix. (In the "Universe" In which we exist, this average is known as "c," a "constant of nature" which is the "speed of light in a vacuum.") At this point there will be a significant change in the balance between the rest of the matrix and the part of the matrix within the surface of the moving entity, this surface will become changed in size and shape. The motion disturbance is no longer primarily dissipated into the matrix at large, but becomes localized at the surface of the involved entity. The balance changes, the "mass" increases. Special

Relativity turns out to have been at least partially correct. The situation changes at the speed of light.

A third factor hidden in the interface between mathematics and reality apparently allows Einsteinian Space-time modelling to give quite accurate predictions. This is in the definition of "Time." In the "Space-Time World" of conventional thinking, "Time" is a reality which somehow came into being at the beginning of "Existence." Actually, time is a convenient method of keeping track of motion in sequence by a measured interval. The hidden factor that apparently makes "Space-Time" modelling work is that "Time" is always referenced to some reproducible cycle in nature. Therefore, "Time" has hidden within it not only the idea of motion, but also the idea of cyclic motion. A unit of time, a second, for instance, can therefore represent a cyclic motion, or motion in a circle, and the expression "sec^2," could stand for the motion content in the volume of a sphere! This insight leads to some interesting interpretations of some of the equations of physics, which, unfortunately, are beyond the scope of the main thrust of this paper.

The "hidden factor in Time." however, makes Space-Time Modelling and Motion in a Matrix Modelling reach much the same conclusions and, at this point, they seem to be essentially equivalent approaches. The Motion in a Matrix users being, perhaps, more cognizant of why their ideas have validity. This "circle/sphere" aspect of time hints at the idea of a spherical oscillator as a basic entity. This latter idea, has become the basis of a variant of Motion in a Matrix Modelling which could be called the "Oscillator Substance Model."

The "Oscillator Substance" model, although developed independently, echoes Max Planck's ideas of dots controlled by oscillators and is the discovery of a chemist who was once trained as an electronic technician.\, who has, within the last year, put together insights from both fields to suggest that there is a very simple model of everything which perhaps would have been close to the "Unified Field Theory" which Einstein spent his life trying to develop but could not. We will come back to this later. First, however, since, we are focused on where Einstein was right, or wrong, we should note the errors that probably doomed Einstein's quest for a "Unified Field" from the start.

It is highly probable that the "First Fatal Error" in Einstein's Unified Field attempt was that he "Threw out the field." That is, he assumed a void, a nothingness, for the "Field" to operate in. The "Second Fatal Error" is one that modern theorists continue to make. They try to set up a unified field theory from the "Four Fundamental Forces." The problem is that the "Four Fundamental Forces" all violate the Law of Forces, "For each and every force there is an equal and opposite force." This law of forces, if examined carefully, can be interpreted to clearly indicate that any "Force" is simply a readjustment of pressures within a "substance." The "Four Fundamental Forces" are either, in two cases (Gravitation and Electro-Magnetism) descriptions of observed phenomena, which are the result of other factors, These kind of "Forces" are known as "Fictional Forces," the best known, and best explained of which is "Centrifugal Force." The other two "Fundamental Forces," the "Strong and Weak Nuclear Forces" appear to be simply imaginative explanations which arise as justification for the idea that neutrons exist, as such, in atomic nuclei. A much simpler explanation appears if one considers atomic nuclei to be electron-proton aggregates in which a neutron has, at the best, potential existence. Einstein's frustration was apparently caused by his operating on sets of erroneous ideas.

In the next couple of paragraphs will appear a statement, which Einstein could have published, a logical outgrowth of Planck's ideas, Which might have kept us from almost a century of what this writer considers "semi-mystical nonsense" in scientific theory.

"Let us assume as a working hypothesis that there exists a 'dot matrix' of separable oscillators. These. in turn are collected to form a 'substance' at its 'triple-point,' of larger separable oscillator entities capable of correlating motion and of separation into the electron-anti-electron set and distortion into neutrona." Amplifications of the ideas in this statement, and corollaries, can give explanations for electrons, protons, the expanding universe, "The Big Bang" and almost any other phenomenon to which it has been applied. It is this working hypothesis which has led to the explanations of "Mass" and "Energy" which have been used throughout this paper.

 Why was Einstein right? When he was, it was almost as much luck as brilliance. The hidden implications of the mathematics which matched reality made the theorizing seem to fit even when the basic ideas erred. Had he gone in a different direction, following up the ideas of Max Planck, he might have reached the same basic conclusions as the "Oscillator Substance Model" which has arrived nearly a century late, and will probably be ignored.

 Post Script: Although this writer finds the Oscillator Substance approach so natural and useful as to feel that it should be common knowledge, the discovery of this possible "Explanation of Everything" is only a few months old. The ideas may wait some years longer for confirmation or disproof as what information published about it, thus far, is almost exclusively on Helium.com. and the discoverer has no professional standing as a member of any research institution or group. It is sad that a young, patent clerk could not have reached these ideas in the early 1900's rather than a very elderly janitor being the discoverer a century later....

 Now, if a certain "String Theorist" were to have happened to have come up with this... Oh, well, we can't have a perfect world!

Einstein was very right to try to explain the workings of reality, it's sad that he made a few key mistakes.

Another article from SciScoop

Four Forces or One Substance?

Conjecture **Tuesday**, January 6, 2009 *by* deanlsinclair

It has been assumed for some years that there are "Four Forces of Nature," as follows: Electromagnetic Force, Gravitational Force, the Strong Nuclear Force and the Weak Nuclear Force.
There is a major problem with all of these "Forces." None of them fit the Law of Forces: " For each and every force there is an equal and opposite force."
A shift of viewpoint may give an explanation for all of this. The Law of Forces could be rephrased to fit a substance, in which case it would read, "Every change in pressure within a substance has a compensatory adjustment within that substance. " If it can be assumed that there be a "Universal Substance," then all true forces are simply pressure adjustments

What in reality are the "Four Forces of Nature" considered in this manner? Electromagnetism and Gravitation are observational Phenomena, what are known as "Fictional Forces," akin to the Centrifugal Force which children on farms used to demonstrate by whirling a milk bucket in a circle. in terms of some "Universal Substance," Electromagnetism would be a description of the results of the interactions of two stable vortex entities, of counter-rotating spins and different sizes
 within the substance.* Gravitation, an apparent "Force of Attraction." would actually be a result of differential pressures in the part of the "Substance" between two "Entities" which are part of the substance, and the rest of the substance, including that part between the two entities.

The two "Nuclear Forces" fall into a different category. Both of them would be constructs developed to justify a conclusion. The Strong Nuclear Force is needed to justify the assumption that neutrons exist, as such within atomic nuclei, despite the fact that under any other known conditions they are unstable with respect to electrons and protons. The "Weak Nuclear Force" has a similar origin from the assumption that electrons and protons can rejoin to neutrons to a an appreciable extent
within conditions to be found in the nuclei of atoms.

The concept of a "Universal Substance" is very old. It was once conceived of as a liquid or gas called an "Aether." This idea of an "Aether" was supposedly destroyed in the early 1900's by the Michelson-Morley Experiment which determined the probable constancy of the Speed of Light in the "Vacuum of Outer Space." By another viewpoint, however, that Constancy of the Speed of Light could just as well have been taken as proof of the existence of a "Substance." The "Substance" simply has different characteristics than had been assumed. While a solid, liquid or a gas, individually, would not be a valid characterisation of the "Substance," something at its "Triple Point" where small disturbances will allow it to react as if it were any one of the three fits very well. Particularly when it is considered that the substance will revert to the "Triple Point State" as soon as possible….

In this view, the "Speed of Light" is the "Maximum Velocity of Information Transfer in the Substance at Rest," or nearly at rest with the disturbance of that information transfer being such that the Substance appears as a "Temporary Solid" during that time of information transfer.

One explanation of this is that the Substance appears to be made up of rotating entities which have a "standard," or average, tangential velocity of "c," the speed of light. (This would be a an angular velocity of "2Pi c " for any readers who are into angular measurement and angular velocities.) Just as belts connecting rotating wheels will transmit information at the "tangential velocity" of the rotating wheels, any set of objects spinning at the same angular velocity are capable of transmitting information at that velocity as a "Carrier Wave."

An analysis made by combining two constants of nature, Planck's Constant, "h," and the Speed of Light, "c," indicates that the "Universal Substance" may be made up of a "family set" of oscillators having the above rotational characteristic and a constant torque (push/pull on one another) of "h/c." *

It may be noted that even Space-Time Theory which appears to not accept the idea of a "Universal Substance," actually, in its very name, not only suggests a Universal Substance, but, also, partially defines it. Mathematical Space is never empty, it is by implication always filled with dots. Time is a measure of sequence which is always referenced to some known cycle. therefore, it can be said that "Space-Time" implies that it operates to describe actions within a substance of tiny units undergoing cyclic motions….

To this writer, the concept of a "Universal Substance" is far more satisfactory than any of the various constructs which have arisen on the basis of the "Four Forces of Nature."

---

*See the following referenced web site for "pages" which pertain.
  http://groups.google.com/group/oscillatorsubstance-theory

Related Posts:

- A Constant's Secrets. A Different Look at Planck's Constant
- An Intertwined Universe?
- Two Energy Expressions Interact?
- Why Einstein was Right

Here is a little fun article that didn't make it to SciScoop

UNIVERSES WITHIN UNIVERSES

The discovery that evaluating Planck's Constant as an angular momentum at the speed of light leads to the mathematical definition of a "constant toque universe of torque, "h/c" (Planck's constant divided by the speed of light) leads to speculation about universes within universes.

The "Universe" defined by the set-equation, { m x r = h/c), mass times radius equals Planck's Constant divided by the speed of light, defines a "family of oscillators" having a constant torque of "h/c" and an average value of $(h/c)^{0.5}$. As a constant on nature divided by another is also a constant of nature. we can say that both of these values are "constants of nature.' So also would be $(h/c)^0$, $(h/c)^{+2}$, $(h/c)^{+1.5}$, etc.

What develops here is the problem that if, $(h/c)^{0.5}$ is the average value of an oscillator family defined by (h/c); (h/c) could be the average value of an oscillator family defined by $(h/c)^2$, and so on. That is, every possible power or root of the ratio could be taken to define another "oscillator family" and hence another "Dimension."

This problem of dimensions within dimensions seems somewhat related to two mysterious numbers of our number system. The numbers "zero," and "one." We say, "Zero is nothing." That, however, is not totally logical, by giving the "number," Zero, a name and a place, it has become something. It can be considered a starting point, a hole, even an entrance into another dimension. In another article dealing with signed numbers,* this writer has discussed the meaning of signed numbers as representing on a three dimensional chart, first a line in a given direction, then a square defined by two directions and third a cube, defined by three dimensions/directions. This writer did not go on to what would have to happen after that. Logically, one can pick a point at which to go on from some given three dimensional cube and move the new origin to that cube, then start the convention over. The first line-up would form a line of these cubes, next change would form a square of these cubes, the next a cube of these cubes and so-on. However, it may be noted that by moving our origin spot, our "zero" to a particular cube, we have made the "dots" in our next diagram be the 3-D cubes of the last diagram. Our new "Zero" is an origin dot, but it is also the cube of the last origin dot. We have taken, in a sense, 1 x l x1x0 and gotten 1^3, which is our new "Zero" but is certainly not "Nothing...." It, in fact, could be considered the "hole" from which are pulling out the "3-D" dots from our previous "dimension. " Another little experiment with "Zero" and "one" also brings in the mysterious number, "Infinity," the "number beyond all numbers," which in practical terms is the next number beyond where we stop counting, for whatever reason. If one forms a "Mobius Strip" by twisting a strip of material one turn and fastening the ends together one obtains a "uni-dimensional infinity," that is, it is possible to go around and around, outside and inside, passing through the same points indefinitely. This strip is both "Zero" and "One" as it is our starting point, and it is one unit.

Now if the one unit be cut by one knife or scissors, one time once around, passing once through the starting point of the cut, what is obtained is still one unit; which, however, is twice as long as the original and twice as twisted. We have run a number of "ones" together to still get "one" but it is no longer the same "one." Continuing by splitting this unit, one obtains, not as might be expected--a longer, more twisted continuous unit--but two intersecting units. As they intersect, they are still "one." Continuing the slitting process produces more and more of the intersecting units, still "one" as they are "attached" to one another by the intersections. We have been always operating on one unit at a time one operation at a time, but we have created many "ones." We reach, "infinity" when we no longer have enough material in each unit to split with our knife. Infinity begins at the point where our "instrumentation" fail.

Since we can set h/c as the basic unit, we could call it the "one" and let h/c=l to develop some sort of "absolute scale." The interesting thing a is that if we do this, since we consider all powers and roots of one as one, we will be defining all of our projected universes as "One."

Going over all of this, the writer feels that at this point, the one thing that should be done with this one article is to write one (.) (:>)
----
*Problems in Mathematics--Signs and Signed Numbers,"
www.sciscoop.com/story.2009/3/19/113736/17

Along the same lines of confusing thought is this older article printed on the Google Group Site, Oscillator/Substance Theory.

THE CONSTANCY OF CONSTANTS

One thing that we seem to depend on is that constants of nature will not change, at least in the time that we are working with them. Going along with this, we can then say that if we combine constants of nature, we will obtain other constants of nature which may or may not bive us insights into what is going on.

One very interesting constant of nature is Planck's constant which relates the concept of "Energy" to the frequency of electromagnetic radiation. This constant, "h," can have the dimensions of grams, centimeters squared per second, or in other words the units associated with "mass," radius, and velocity, the units of "action" or "angular momentum. We can write then that "h" as a constant can be considered as m x r x v = h bing a defining equation. This is of the form of three unknowns equaling a constant, mathematically, that is, "xyz=K." We also notice that, if we divide out the factor,"r." we have another rather familiar looking equation, mv=a constant. This is the familiar equation for momentum, "p." Therefore, h/r is the same as momentum. We may note, however, that there are two other ways we can change things to get something that looks auspiciously like a momentum. We could divide out mass or velocity and get an equation of the same form. Maybe we have just discovered the rationale of the rule that in collisions, momentum is conserved. In a three factors equal to a constant, if one factor changes, the other two will change in such a way as to keep the relationship the same.

Another way of looking at this is to note that since all three are related to the concept of angular momentum, all three factors are related to some central point. The radius, r, is a distance from the center, "v," would be the velocity of a unit moving at that distance, and the mass would be the balanced pressure at that distance that kept the motion constant.,

At this point, let us introduce another constant of nature, the speed of light, "c," which, since it is the velocity of information transfer by electromagnetic radiation, has to be the average velocity, in any given direction of any moving entity..

Evaluating mxvxr=h at "c" by dividing out "c" we get m x r =h/c.   This is the equation for a torque, mxr. so "h/r," is the "torque constant of nature." Since the absolute values of m and r are interchangeable, we have defined a set of rotors which can have an unlimited number of pairs of values that will fit "m x r = h/c." We may also say that this set of rotors would also define an oscillator family set, m x r = h/c.

It is not too great a jump to say that since this seems to be a constant of nature that our Universe itself would be within the most powerful low-frequency member of this family, with all other units "keyed" to "harmonics" within that same oscillator. Going the other way, looking at certain other phenomena, such a electron-anti-electron pair production, and e-, e+ annihilation, moves the concept to the idea of these oscillators being either "full-cavity" oscillators, which would account of r a Universe/Anti-Universe pair, or a positron-negatron, "zerotron" parent unit, or being vortex "half-oscillators" which would account for the opposite spin characteristics of electrons and anti-electrons and can be extended to account for the structure of all matter as being composed from electrons and protons.
We can go on, we can divide h by c^2. to get a constant. We can divide c^3 by h to get a constant. In fact, we can manipulate any set that we wish of h and c to get possible constants of nature. The one mentioned above of "c^3"/h, has the dimensions of acceleration per gram, which if changed by multiplying out grams as one gram, gives a huge number as a force. Perhaps the force that would be felt, is being felt with out our realizing it, on every gram total of units of existence.... Some of these ideas of the 'Principle of the combinability of constants as clues to the structure of everything," can be explored further.

If anyone has read this far, how about telling what "h/c^2" might mean? How about some of the other combinations?

Lets switch to a couple of math. articles that got to SciScoop, but I seem to have lost the capacity to reprint from there.so back to my files. Looks like I got this copied from SciScoop before.

Problems in Mathematics--Signs and Signed Numbers	
By deanlsinclair, Section Commentary Posted on Thu Mar 19, 2009 at 07:37:36 AM PST	+Hotlist
The various meanings associated with the positive and negative signs in the simple mathematical processes produces some interesting problems with signed numbers.	

The signs, + , and, - , are used throughout mathematics and physics in a number of ways,and with several meanings which are often not carefully checked. The plus or positive sign has its initial use in addition in the sense of increasing a pile, of no particular dimensions, by a certain amount described by a

counting number written after it. The negative sign represents the opposite operation of removing a certain specified amount.

In Physics, the positive sign represents a "charge" associated with the proton, the negative sign represents and "opposite" charge associated with the electron and other species having a characteristic in common with the electron. [This writer suspects that this characteristic is a counter-clockwise spin.] In this usage, the signs do not represent reversed operations but characteristics which are considered opposites.

A third usage shows up in mathematics where the signs are associated with counting numbers to form sets of "signed numbers" which seem to be able to be added, subtracted, multiplied and divided like counting numbers. However, this turns out to have problems when one does multiplication and division processes. What is overlooked is that the addition of the sign to a number gives it both a magnitude and a direction. Something having both magnitude and direction is not a true number, it is what is called a "vector." In this case, one might call it an "operator" for the sign associated tells the direction of motion of the next operation in which the number may be involved.

The signed number may represent movement away from a zero point, a line, a plane, a three dimensional figure formed of planes or some "higher order figure" depending on where it occurs in a sequence of operations. Signed numbers are handled according to a convention wherein the positive sign is considered as being to the right of an origin, upward from an origin or forward from an origin, and if one multiplies three positive signed numbers together, say plus two times plus two times plus 2, (+2 x + 2 x +2) what one has really described is moving two units to the right of the zero point, moving this "two-units-line" upward to form a square, then moving this square two units forward to create a cube which is situated to the right, above and in front of the origin point. This "eight-cubic-units entity is called a positive volume because we say + x + x + = + as we consider that the + sign represents travel " in the same direction" while the negative sign represents travel in the opposite direction. This signed unit, like the line and the square, has a direction associated with it which would be at right angles to the last upward motion. Labeling this unit as positive continues the vector content but is not truly in accord with the "reversal idea upon which it is based, for, as we have seen above, each operation represents a change in direction, but of 90 degrees, not a reversal. If we go, + 2, +2, -2, in our sequence of operations, we will go to the right first, up second, and back from the "center-plane" third to form another eight unit volume which will be above, to the right, but behind the point of origin. This will be considered a "negative volume" purely by convention as it has one negative sign associated, however, this convention preserves the vector designation.

As the positive numbers are associated with "positive values," right--as in handedness, upward--toward the Heavens, and forward--"progress." Negative numbers are associated with the reverse, left--"sinister" or left-handed, downward, and backward. The operation, -2 x -2 x -2 , would create a cube, which was to the left of the origin point, below the 'origin line," aka, the "x-axis" and behind the "origin plane," aka, the "xy-plane." Note that the first square formed would be considered a "positive number as it is "minus x minus = plus" but the third operation, adding another direction considered "minus" labels the resulting volume as a "negative number volume."

Summing the above, a signed number represents a line, two signed numbers multiplied together represent a plane and three signed numbers multiplied together represent a volume and, in the order of the multiplication process, will determine what plane or volume is described. The operation, " +2 x -2 " represents, by the conventions used, a square which is to the right and down from the origin. while the reverse operation, "-2 x +2" forms the representation of a square which is to the left and above the origin.

It can be seen then that while the operation, 1 x l x l ,as counting numbers still represents the original one. Plus-one times Plus-one times Plus-one represents one whole, but it is one whole cube, one length unit to a side, not the original one line....! Similarly, it can be seen that the cube root of eight as a counting number is simply the number two. The cube root of +8, as a "vector cube" has the "absolute value" of 2 but this two can be either a positive or negative vector depending on which of the "generating sets" it belongs to and the order in which it falls in the set. A positive-volume-vector, +8, can be generated by any of four sets of three "signed twos." These sets are as follows: {+2, +2, +2}, {-2, -2, +2}, {-2, +2, -2}, or {+2, -2, -2}. A negative-volume-vector,-8, can be generated by any one of the sequenced-operation sets {-2, -2, -2}, {-2, +2,+2}, (+2, +2,-2} or {+2,-2,+2}.

Assigning a "signed-root" to a signed number, is therefore a difficult and tricky business which would actually require a knowledge of the history of the signed number in question! That is we would have to see the graphic representation of the particular cube with which the -8 is associated. It is no wonder that the mathematicians seem to ignore "odd-number" roots of signed numbers and consider that the square root of minus one is "plus or minus 'i,' an imaginary number," which truly it is. In the most basic unit of "minus one" we are considering a line vector of a unit length. How does one take a root of a line vector "running backwards?" Actually the root would have absolute dimension of one, either plus or minus, as one would have to be speaking of the "second-order-vector-square" which can be generated by either of the sets, {+1, -1} or {-1, +l}. Mathematicians have no trouble with saying that the square root of +1 is plus or minus one as it is generated by the two sets, {+1, +1} and {-1, -1}. These are sets within which the internal values appear to be identical to one another. As one can see from previous discussions that the internal elements are not identical but represent different directions of the vector depending on their position in the sequence. One of the square elements would be to the right and above the origin. The other, below and to the left. In saying that the square root of Plus-one is "Plus or minus one." we are actually talking about the square roots of two different "ones...."

The use of the two signs with different meanings of operation, reversal, or direction causes some interesting problems in understanding mathematics.

The following is only a slight variant on the above article, I don't know how both got published under different topics in SciScoop as they are essentially identical. If you've read the above article, the next will only be review....

On Signed Numbers

The signs, + , and, - , are used throughout mathematics and physics in a number of ways,and with several meanings which are often not carefully checked. The plus or positive sign has its initial use in addition in the sense of increasing a pile, of no particular dimensions, by a certain amount described by a counting number written after it. The negative sign represents the opposite operation of removing a certain specified amount.

In Physics, the positive sign represents a "charge" associated with the proton, the negative sign represents and "opposite" charge associated with the electron and other species having a characteristic in common with the electron. [This writer suspects that this characteristic is a counter-clockwise spin.] In this usage, the signs do not represent reversed operations but characteristics which are considered opposites.

A third usage shows up in mathematics where the signs are associated with counting numbers to form sets of "signed numbers" which seem to be able to be added, subtracted, multiplied and divided like counting numbers. However, this turns out to have problems when one does multiplication and division processes. What is overlooked is that the addition of the sign to a number gives it both a magnitude and a direction. Something having both magnitude and direction is not a true number, it is what is called a "vector."

The signed number may represent movement away from a zero point, a line, a plane, a three dimensional figure formed of planes or some "higher order figure" depending on where it occurs in a sequence of operations. Signed numbers are handled according to a convention wherein the positive sign is considered as being to the right of an origin, upward from an origin or forward from an origin, and if one multiplies three positive signed numbers together, say plus two times plus two times plus 2, (+2 x + 2 x +2) what one has really described is moving two units to the right of the zero point, moving this "two-units-line" upward to form a square, then moving this square two units forward to create a cube which is situated to the right, above and in front of the origin point.   This "eight-cubic-units entity is called a positive volume because we say + x + x = + as we consider that the + sign represents travel " in the same direction" while the negative sign represents travel in the opposite direction.   This signed unit, like the line and the square, has a direction associated with it which would be at right angles to the last upward motion. Labeling this unit as positive continues the vector content but is truly in accord with the "reversal idea upon which it is based, for, as we have seen above, each operation represents a change in direction, but of 90 degrees, not a reversal. If we go, + 2, +2, -2, in our sequence of operations, we will go to the right first, up second, and back from the "center-plane" third to form another eight unit volume which will be above, to the right, but behind the point of origin. This will be considered a "negative volume" purely by convention as it has one negative sign associated, however, this convention does preserve the vector designation by the conventions observed.

As the positive numbers are associated with "positive values," right--as in handedness, upward--toward the Heavens, and forward--"progress." Negative numbers are associated with the reverse, left--"sinister" or left-handed, downward, and backward. The operation, - 2 x -2 x -2 , would create a cube, which was to the left of the origin point, below the 'origin line," aka, the "x-axis" and behind the "origin plane," aka, the "xy-plane." Note that the first square formed would be considered a "positive number as it is "minus x minus = plus" but the third operation, adding another direction considered "minus" labels the resulting volume as a "negative number volume."

Summing the above, a signed number represents a line, two signed numbers multiplied together represent a plane and three signed numbers multiplied together represent a volume and, in the order of the multiplication process will determine what plane or volume is described. The operation, " +2 x -2 " represents, by the conventions used, a square which is to the right and down from the origin. while the reverse operation, "-2 x +2" forms the representation of a square which is to the left and above the origin.

It can be seen then that while 1 x l x l as counting numbers still represents the original one. Plus-one times Plus-one times Plus one represents one whole, but it is one whole cube, one length to a side, not one line....!

Similarly, it can be seen that the cube root of eight as a counting number is simply the number two. The "cube root of +8, as a "vector cube" has the "absolute value" of 2 but this two can be either a positive or negative vector depending on which of the "generating sets" it belongs to and the order in which it falls in the set. A positive-volume-vector, +8, can be generated by any of four sets of three "signed twos." These sets are as follows: {+2, +2, +2}, {-2, -2, +2}, {-2, +2, -2}, or {+2, -2, -2}. A negative-volume-vector,-8," can be generated by any one of the sequenced-operation sets {-2, -2, -2}, {-2, +2,+2}, (+2,

+2,-2} or {+2,-2,+2}. Assigning a "signed-root" to a signed number, is therefore a difficult and tricky business which would actually require a knowledge of the history of the signed number in question! It is no wonder that the mathematicians seem to ignore "odd-number" roots of signed numbers and consider that the square root of minus one is "plus or minus 'i,' an imaginary number, which truly it is for in the most basic unit of "minus one" we are considering a line vector of a unit length and how does one take a root of a line vector "running backwards?" Actually the root would have absolute dimension of one, either plus or minus, as one would have to be speaking of the "second-order-vector-square" which can be generated by either of the sets,   {+1, -1} or {-1, +l}. Mathematicians have no trouble with saying that the square root of +1 is plus or minus one as it is generated by the two sets, {+1, +1} and {-1, -1}, sets within which the internal values appear to be identical to one another. As one can see from previous discussions that the internal elements are not identical but represent different directions of the vector depending on their position in the sequence.

The use of the two signs with different meanings of operation, reversal, or direction causes some interesting problems in understanding mathematics.

The following little article is a follow-up of the above one.

Roots and Directed Numbers

In working with numbers one often works with fractional exponents, "roots." This works well when one is working with "absolute values, " unsigned numbers; however, it runs into complications as soon as one starts to operate with "signed" numbers. The problem probably arises from the inherent fact that absolute number values do not have a directed motion automatically assigned to them. A positive number is by convention associated--usually--with a motion upward, to the right, or forward, with a negative number associated with motion downward, to the left, or backward. When we take the square root of one, unsigned, we're are talking about a number which multiplied by itself gives the original number, one. When we take the square root of 4, we realize that it is the number two, when we place two units down twice we get four. When we are working with signed numbers we have a different situation. If we are taking the square root of +4, we are actually asking the question, "What is the directed side length of a square which we consider to have the area "Positive Four" when we operate according to the conventions associated with signed numbers?" By those conventions we can see that both +2 times +2 and -2 times -2 fit this criterion, so we say, quite correctly that the square root of +4 is either +2 or --2, Perhaps we would, however, have been more accurate in saying that there are two sets of square roots to the number, +4, the set, +2,+2, {+2, =2} and the set ,{-2,-2}.

The reason that this last was said will become clear when we discuss the situation for the "square root of -4." Let us analyze this problem as we did above. The question we are asking is what two directed numbers will produce an area which by our conventions of directed numbers will be assigned a value of -4?" This occurs again in two cases, producing two sets, {+2, -2} and {-2,+2} . As these are directed numbers the set, {+2,-2} is not identical to the set {-2, +2} as they represent opposite directions of sequential motion. With the "Positive Area" we find that we can create what we call a positive area by going in a positive direction then turning in another positive direction, or going in a negative direction and then turning in a negative direction again. For a negative area we can start out in a positive direction, then "turn negative" or start in a negative direction and "turn positive." By this analysis, the square root of "Negative One" is not an imaginary number but can be said to be not as in the other case, "Plus or Minus One" implying each "operating" on itself, but "Plus and minus one" the two operating on each other. The concept of imaginary numbers arises because of the ignoring of this fact of the directed action factor inherent to signed numbers.

This can, of course, be extended to higher roots. For the cube root of +8, one may write the sets, {+2, +2, +2}, {-2,-2,+2}, {+2, -2, -2}, and {-2,+2,-2}. Noting four sets that can be considered the "cube root" of +8. A similar group of 4 sets represents the "cube root" of -8. A fourth root would presumably continue the pattern developing eight sets of 4 units each. This is left to be proven, or disproven by the reader.

While the idea of imaginary numbers as successive even roots of "Minus One," is an interesting concept, but by the above analysis appears to be based on a misunderstanding of the significance of signed numbers. The use of a signed number indicates a motion in a direction and can be considered to define a "dimension."

From simple mathematical ideas to another view of electrons.

SET NOTATION AND ATOMIC STRUCTURE

The equivalence of electrons in atoms is more or less implied in Oscillator/Substance theory and is explored somewhat in another page. Here is a little different slant.

Electrons in atoms may be considered in a set theory manner as consisting of one set which tends to subdivide into two other sets. One of these sets, the "nuclear electrons" have been considered since the 1930's as being bound to protons to form neutrons. Another possibility is that they have motion patterns, "orbitals" that are analogs of the orbitals that are assigned to the extra-nuclear electrons or to the orbitals which are written in molecular orbital theory for molecules.

If we take the first idea and use a simple example, what can be called the "Iso-3,3-set members, Tritium and Helium 3," we can write, using the form, N-1s2; 2s2, etc.,for nuclear electrons and a corresponding, E-1s2;2s2, etc. notation for the "extra-nuclear electrons, we would have the following sets:
    {e-H3} = {N-1s2} U {E-1s1} and {e-He3}= {N-1s1} U {E-1s2}. This would read, " the set of electrons in Hydrogen 3 is composed of the set of 2 electrons in a "ls" type orbital in the nucleus and and a 1s electron that is not in a nuclear orbital. The situation for Helium 3 is a union set of one electron in the nuclear orbital and two in an outer orbital."

The Union set notation implies that all electrons belong to the entire unit....

Protons can also be supposed to have some degree of motion which could perhaps be described in the same way. In that case we could write for protons in the above cases, for both units, {P-1s2, 2s1} and an entire description for Tritium or He3 could be done by writing the foregoing set as a union set with the electron set to describe the entire atom.

Whether any of the above will spark ideas for research or computer modelling is unknown at this point. There is one paper, published on www.helium.com as "A Guide to Helium II" which suggests that the super fluid characteristics of HeliumII could possibly be due to the change of the nucleus from an tetrahedral to a square planar configuration.

So why not reprint the article, "A Guide to Helium II" here? Sure, why not?

A GUIDE TO HELIUM II
Dean L. Sinclair

Helium 4 has two liquid forms. Helium I, which is a "normal liquid," and Helium II, "He4II," to which it sharply converts at 2.174K. This strange material expands on cooling, has tremendous heat conductivity, a strange viscosity that it will climb up the sides of the container and is the only liquid known which can not be solidified by cooling at ordinary pressures, but will solidify under pressure.

With increased pressure the volume of the solid can be decreased more than 30%. A very strange substance indeed, for which there seems to be no explanation in the literature.

There is a possible explanation which would need experimental verification by someone having access to the situation to be able to check it out. The postulated explanation is as follows: At 2.417K. ("K" in this context means "degrees Kelvin, i.e. degrees above Absolute Zero, the He4 nuclei, up to this point consisting of four protons in a tetrahedral array (for the moment we are ignoring the electrons bonding them together) lose the internal vibrational energy necessary to maintain their equal spacing in 3D space and collapse into a 2D, square-planar ground state, with bonding electrons now assuming positions above and below the plane of the "square." This type of activity is unknown for any other nucleus, and would be impossible for almost any other array but corresponds roughly to carbon as a diamond atom changing to carbon of the type found in graphite.

A liquid made up of square-planar atoms would arrange itself automatically into flat plate arrays which could slide over one another and conduct heat easily along the plane of the plates. This would account for the strange viscosity and the heat conduction.

It does not seem to have been noted that, while Helium I, "He4I," would be expected to be an insulator, there is a strong possibility that He4II would be a good conductor of electricity. Solid Helium, which would presumably be "Solid He4II" might add additional evidence in support of the above theory or be able to refute it. Normal Helium would be expected to be a tetrahedral nucleus, resulting in a "ball" atom which would have a "cubic-close-packed" crystal structure. If the above speculations are correct, solid He4II would have a square-planar crystalline structure. The cubic structure would be an insulating dielectric. The structure postulated above would be very likely to show conduction in plane of the square plates which would vary with pressure at right angles to that plane.

A thin film of either of the solid or liquid forms of He4II could very well show colored patterns when viewed with "crossed Polaroids." (This is a technique wherein light is polarized, passed through a substance, then passed through another polarizing lens, films having varying structures within will produce beautiful patterns which can be varied by rotating one of the polarizing lenses.)

Liquid and solid He4II may have some very interesting implications for theoretical chemistry and some very valuable undiscovered uses.

The above ideas need experimental verification or refutation. If this writer's speculations have merit, someone who could verify them would be able to write some very interesting papers. Anyone having a friend working in the field of cryoscopy with liquid Helium is strongly urged to refer their friend to this article.

_____ Note: Wouldn't it be great fun if a short article on Helium II published on www.helium.com could spark some advances in chemical theory?

Is anyone curious as to how all this theorizing started? Well, the Helium article, "Motion in a Matrix...." posted in April of 2007 was the first Internet paper. Somewhat simple and naive, it can stand republishing here;

Motion in a matrix as a new model of the physical universe

by **Dean L. Sinclair**

A combination of a basic idea about communication and an analysis of motion is seen as an indication that a possible model of existence is "motion in a matrix." This is in the first section of this tripartite essay which runs approximately 3400 words.
In the second section, energy expressions are equated to estimate some possible characteristics of the postulated matrix.

The third section explores the implications of the concept of momentum with respect to this type of model.

A model for our physical universe considering motions in a matrix as being fundamental, may be usable to explain most , if not all of our present knowledge. That the speed of light is a constant of nature which can be considered to be the limiting velocity of an information carrier-wave, and that all information moved by other means, can be considered as being moved from particle to particle by some sort of "wave motion," suggests that electromagnetic radiation,light,is a transverse wave motion in a three-dimensional matrix analogous to a solid. Quantization of every known interaction, also implies an underlying, regular structure, a "Matrix."

"Motion" of some sort is a "given" in all observations/experiences. The fact that motions can be analyzed in two categories, motions related to a point or points, and somehow "attached" to the point or points, and motions along a line, fits well with the known Mass/Energy duality. Mass is considered as static, i.e. it could conceptually be related to motion associated with a point, while Energy is noted when motion along a vector is changed in direction or impeded. This somewhat reverses the concept of Energy from "That which moves Mass, " to "That which is observed when Mass is moved." Mass, then, may be considered to be a characteristic of a point-centered motion in a matrix, a characteristic of the spinning of a "3D-Vortex."
We, therefore, can conceptually consider all matter as being made up of combinations of two variants on two fundamental, stable, "Vortex Particles" having opposite spin-tumble orientations. These would be what we know as the electron and positron.
[The proton can be considered as a very large vortex formed by partially inelastic collisions

between positrons, resulting in "Exploded Positrons," thus explaining the absence of positrons in our "normal world." They are "hiding" in the guise of the proton. Vortices of opposite spin/tumble can be expected to attract, and if of the same size could mutually annihilate.

Well, this ran to eight pages on Helium. After reprinting the above first page, it looks like it will be more efficient to go to our own files and print out the version that is there!!

MOTION IN A MATRIX, REVISITED.
THIS POST INCLUDES MOST OF THE EARLIER ARTICLES ON THE SAME SUBJECT. IT IS ESSENTIALLY A DUPLICATE OF THE ARTICLE POSTED ON www.helium.com AS A "LEAP-FROG" OF THE EARLIER ARTICLE.
    A combination of a basic idea about communication and an analysis of motion is seen as an indication that a possible
    model of existence is "motion in a matrix. This is in the first section of this tripartite essay which runs some 4500 words.
    In the second section, energy expressions are equated to estimate some possible characteristics of the postulated matrix.
    The third section explores the implications of the concept of momentum with respect to this type of model.

A model for our physical universe considering motions in a matrix as being fundamental, may be usable to explain most , if not all of our
present knowledge.

That the speed of light is a constant of nature which can be considered to be the limiting velocity of an information
carrier-wave, and that all information moved by other means, can be considered as being moved from particle to particle by some sort of
"wave motion," suggests that electromagnetic radiation, light, is a transverse wave motion in a three-dimensional matrix analogous to a
solid.

Quantization of every known interaction, also implies an underlying, regular structure, a "Matrix."

  "Motion" of some sort is a "given" in all observations/experiences. The fact that motions can be analyzed in two categories, motions
related to a point or points, and somehow "attached" to the point or points, and motions along a line, fits well with the known
Mass/Energy duality. Mass is considered as static, i.e. it could conceptually be related to motion associated with a point, while
Energy is noted when motion along a vector is changed in direction or impeded. This somewhat reverses the concept of Energy from "That
which moves Mass, " to "That which is observed when Mass is moved." Mass, then, may be considered to be a characteristic of a point
centered motion in a matrix, a characteristic of the spinning of a "3D-Vortex." We, therefore, can conceptually consider all matter as

being made up of combinations of two variants on two fundamental, stable, "Vortex Particles" having opposite spin-tumble
orientations. These would be what we know as the electron and positron.

[The proton can be considered as a very large (turns out on further collocations that the proton vortex is apparently a compression rather than an explosion. Note added, July 2009) 'vortex formed by partially inelastic collisions between positrons, resulting in Exploded Positrons.; thus explaining the absence of positrons in our "normal world." They are "hiding" in the guise of the Proton. Vortices of opposite spin/tumble can be expected to attract, and if of the same size could mutually annihilate. If the vortices are very different size/motion characteristics, the attraction will still be there but the destructive potential is not. The neutron can be considered as an
electron-proton combination, with the electron spinning within the proton. (And outside and through...July 2009)

The mass/energy interconversion would imply that any given vortex would have a certain amount of motion associated with it, as linear motion increased, point-centered motion would decrease, and vice versa. Applying this to electron orbitals, the electron at its farthermost distance from a nucleus would have it maximum point-centered motion, "mass," and it least linear motion, "energy." At the center of its motion passing near or through the nucleus it would have its minimum mass and maximum energy.

Now, one should look at the Space/Time dimensionality, Space can be considered as an inherent characteristic of the matrix. Time is always measured by consecutive, changing motion and can, therefore, be considered as just that, a measure of consecutive motion, referenced to some reproducible cycle.

What are some of the other results of this view?
The fact that light is considered a "mass-less particle" is no problem in this model. It is a linear disturbance in the matrix.
A positive charge would be associated with one spin/tumble orientation. A negative charge with the reverse orientation. Gravitation would be a result of the strained Matrix trying to readjust.
Motion of the electron vortices in atoms and molecules would fit best with the "3-D Pendulum" conception of electron orbitals. (More on this later.)
This whole area needs far more work. Here are some of the questions which come to mind.
What are the "dots" of the matrix?
Neutrinos in a ground vibrational state might fit. (This would explain why a neutrino would travel at close to the speed of light. A "freed-up-neutrino" would pass through the matrix in the same way that a pendulum ball hitting a string of like-sized balls will bounce another pendulum ball off the other side. It wouldn't be the same neutrino; but, who could tell the difference?)
What is the spacing?
Is this spacing implied in some of the constants of nature, just as the constant speed of light implies the existence of the matrix? (This will be explored more fully later.)
Can the tremendous amount of physical science data that is now extant be fitted to this model?

This model does nothing to solve the question of existence, but does seem to cut down on the number of unknowns with which one must deal.

W e shall make an attempt here to evaluate the spacing and vibrational frequency of the "dots" that are proposed as making up the universal matrix. We can attempt to do so by starting with two constants of nature. These are the speed of light in a "vacuum" and Planck's Constant, "h," which relates energy to vibrational frequency.

The postulation of a matrix has the inherent prediction a high-frequency cut-off point for information/energy transfer. it can be postulated that this cut-off frequency would be at the wave length of the spacing of the "dots." It also would be at the vibrational frequency of the dots, and at the highest energy "quantum" that could be carried by a carrier wave passing through the matrix. The Kinetic Energy of this quantum--all energy being considered, at least for now, as actually being "kinetic"--would be $1/2\, mc^2$ wherein "c" is the speed of light. (The actual value of a quantum that could pass through would technically be infinitesimally less than this value, but we can't measure that close.) This Kinetic Energy could also be measured by the expression, $E=hv$, where "h," is the afore mentioned Planck's Constant, with a value of
6.63×10^{-27} erg. sec. , and "v" represents the vibrational frequency.

We can equate these two expressions to obtain $1/2\, m\, c^2 = hv$. This we can rearrange to find the numerical value of "v." In this form we have $v = mc^2/2h$. Plugging in the values of 3×10^5 cm./sec. for "c" and 6.63×10^{-27} erg.sec. for "f, " and doing the calculations, we obtain a value of 6.9×10^{36} cycles per second for the vibrational frequency. (This calculation is independent of mass, the mass units cancel out.) Since the velocity divided by vibrational frequency would give the wave-length, the above figure divided into the speed of light gives a spacing of the units in our matrix as approximately, 4.3×10^{-32} centimeters.

This can be compared with the Planck's distance derived by Max Planck by a combination of the Gravitational Constant, a variation of his constant known as Dirac's constant and the speed of light. This number, which is sometimes called "The infinitesimal black hole from which no light could escape," is $1.6l \times 10^{-35}$ meters or 1.61×10^{-33} cm., approximately 1/26th of the size that we calculated for a possible spacing. As the calculation implies that the number could possibly be the diameter of a circle or a sphere, we may speculate that Planck's number represents the diameter of one of the "vibrating dots" in our matrix. (For much more technical information about Planck's number and the "Natural Scale" based upon it, check Wikipedia.org.)

Now what about the mass of our dots? The amount of mass is immaterial in the above calculation as the mass units cancel out. The mass of our dots can have any value and the calculations would remain the same as long as we're talking about the same amount of mass. For the moment we can presume our dots to have an "infinitesimal" mass as close to zero as we wish it to be. The interesting result of this is that there is no apparent limit to the size of the mass/energy unit which can move through the matrix. It appears that the Mass/Energy could be considered together as either a "measure of motion" in the matrix, or a measure of the instantaneous involvement of the matrix in this particular set of motions.

It can be seen that what we are talking about as the basis of everything seems to be an unimaginably small unit moving back and forth at a extraordinary rate in what we would consider an infinitesimally small space--and with an incredibly huge number of them in every cubic centimeter of space. Assuming a regular cubic structure, we would have some 2×10^{96} dots per cubic centimeter of a "vacuum."
we seem, in a way, to have a considerable simplification from the Standard Model developed in 1970-73 with its bewildering array of fundamental particles.(see the Wikipedia article, "Standard Model.") We still, however, are left with a situation of great complexity. It is highly likely that we shall discover that our "points" have some sort of structure,

including a "rest mass," angular rotation, spin/tumble. our dots may turn out to be actually sets of some sort. We have moved some of our problems of structure to a very micro scale, but, they are still there. (How does the old poem go, "...and smaller fleas to bite 'em, ad infinitum?")

Very long wave length vibrations in the matrix would set up what would amount to membranes and cutting planes through membranes would produce "strings." Brane and String theories may not be totally incompatible with this model. This is especially true when one realizes that the 10 dimensions of the Brane and String Theories--and the 9 tensors mentioned by Einstein--can be considered as ways to "I.D." a point where there is no fixed reference. (One might say that it is a way to locate yourself, in theory, when you have not the slightest idea where you are.) If a fixed reference set of axes were to be found, the 10 dimensions, and the equivalent 9 tensors, could be reduced to three dimensions of "where" and one of "when."
That the gravitational forces would be most likely a result of the matrix, as an entirety. trying to return to an unstressed state may well be consistent with the long held belief that gravitation was due to very long wave motions in "Space/Time." It seems rather obvious that Space/Time is simply another name for our dot matrix. Long wave motions some of which could have wave lengths of many light years are certainly possible. (O.K., O.K., probable, essentially certain.) The "Long Wave" explanation of gravitation may be quite valid.

In considering the involvement of momentum and energy in Motion in a Matrix ideas, the writer has come to the conclusion that, for stability, a system will have a total constant energy of two types which may be called "Kinetic Energy," associated with a vector, and "Static Energy" associated with a point and that "Mass" must be considered as a "Velocity" in the opposite direction of the normally considered motion vector. A definite possibility is that it is a measure of the vector sum of all of the rotational energy vectors of the involved "dots" in the opposite direction of the linear motion of the particle. (This involvement of momentum will be discussed later.)
It may also be noted that Planck's constant has the dimensions of "Action" or "Angular Momentum." This has long been known; but, previously has been given no interpretation. It can be now interpreted, in our model, to suggest that "Energy" is related to the angular momentum/action involved with all of the points involved in the system which we happen to be investigating.

Since we have mentioned momentum, we shall go ahead with integrating momentum into orbital theory in terms of our model.

For many years there has been questions as to how particles could have stable orbits. At the same time, it has been long known that the product of mass times velocity known as momentum, "p," seems to be a constant for any given system. Recently it was noted that momentum, considered as a mathematical expression, could be considered as either an integration or a derivative. That is speaking in the sense of differential and integral calculus. As an integral it is the integral of force over time, in other words it sums up the total effect of all forces that have ever operated against the object on question. As a differential, it can be considered as either the instantaneous rate of change of velocity with respect to time, with mass being considered constant. It may also be considered the instantaneous rate of change of mass if velocity is held constant. This latter does not seem to have been explored.

In mathematical terms the expression, mv, can be integrated as if it were, $mv \times dv/dt$, to give the familiar Kinetic Energy expression. $KE = 1/2\, mv^2$. Alternatively, it can be integrated as if it were vm x dm/dt to give another expression of the same form--which apparently has been overlooked previously, E=1/2 vm^2. This expression we shall call, by contrast to Kinetic Energy, "Static Energy." These appear to correspond respectively to the "vector characteristic" of "Energy," i.e., Kinetic Energy, and the association of motion centered on a point which is postulated in this Motion in a Matrix Model.

Assuming that both integrations are valid, and knowing that Mass and Kinetic Energy have some sort of inter-convertibility, we can write, Total Energy ("TE"), equals Kinetic Energy plus Static Energy and assume that for a stable situation, "TE" is equal to some constant, "C." So, for a stable "orbit." of any kind we can write: TE=SE + KE, or C=1/2 vm^2 + 1/2 mv^2. We can discard the "1/2" and write something like "C= vm^2 + mv^2. We can see that this adds up to the equation, C = pv + pm , i.e. the total energy content of the system is equal to the momentum times the velocity plus the momentum times the mass.

For a stable system, this energy content is, presumably, a constant. If the velocity in a given direction goes up, the mass goes down. At the point of greatest velocity in a stable system the mass would be the least, at the point of least velocity, the mass would be the greatest. This prediction from the mathematics is being done here without checking any of the literature, but the writer is willing to bet that the measured mass of planets varies from perihelion to aphelion in exactly the way mentioned above.

This fits also with the idea of a "3-D pendulum" model for electron orbitals wherein the electrons would pass through and about the nuclei in a three-dimensional analog of a pendulum wherein they would have their greatest mass at the farthest extent of their motion and the greatest speed at the center of the atom. This could account, at least to some extent, to the "solid" feel of matter.

In the case of " Black Holes," there apparently arises the situation wherein "v" reaches a directional limit and any interaction which would have increased the velocity in that direction goes instead to increase "mass." or, perhaps more correctly, an interaction which would have increased the Kinetic Energy increases the Static Energy instead. The idea is that a "Black Hole" would occur at any time when some object reached the speed of light along any one vector. As any further acceleration along that vector could not be compensated for, the mass would have to increase causing a condition of instability with radiation necessary until stable states were reached by the components of the interaction. In this view, a "Black Hole" is not some very mysterious gravitational sink, it is simply some object which has reached the speed of light along some vector.

 [Although Einstein's Relativity limit of "No speed greater than light" cannot apply to relative velocities, there is, in the Motion in a Matrix model, a basic notation that the fastest transport speed in the matrix is "c" for either information or energy. This would also include "Matter" as matter is considered as an indication of involved motion in the matrix. Therefore, Einstein seems to have been intuitively correct that no body will accelerate beyond the speed of light. Attempted acceleration will simply increase the "Static Energy" which is related to what we measure as "Mass."]

An interesting situation develops if one equates the two energy expressions. One finds that "m" = "v" in this case. If instead of equating the two expressions, one assumes a constant of zero, then the logical result is that m= -v; i.e. if the mass were exactly equal to the velocity, the total energy content of the system would be zero Both of these conclusions lead to the idea that "mass," which we consider as being point-centered, may be a measure of the vector component of the angular velocities of the spinning "points" which is directed in the opposite direction of the vector of travel of the entire "body."

While this article is talking about the integration of the momentum definition equation, it may be noted that the same thing can be done to the Force equals Mass times Acceleration equation, F = ma, to obtain analogous expressions to those noted above, as the summation of force over time is known to be momentum, both $1/2\, am^2$ and $1/2\, ma^2$ have to be expressions of momentum and presumably can be summed as was energy, giving a similar conclusion to the one postulated above for energy. The conclusion being that, for a stable system, momentum summation and energy summation will both be constants.

All of this seems consistent with the idea implicit to the Motion in a Matrix Model that a stable system would most likely be one which involved a fixed number of units of the matrix, or at least a certain fixed amount of "action" in the system. It is to be noted that Planck's constant which relates Energy to Frequency has the dimensions of "Action" or "Angular Momentum" implying that the two terms are synonymous.

Final note: This entire concept is, thus far, the work of one person, using as a primary tool a quite ancient "biological computer." If others become involved, corrections, additions and amendments should arise. It may be noted that the first publication of a version of this paper was about April 1, 2007; however, this is not an April Fools Joke; but, rather, an attempt to add another possible tool for the understanding of the Existance of which we are a part. Only time will tell whether it will be accepted as having value or simply added to the waste bin of discarded, "obvious pseudo-science," as are so many new ideas; e.g. those associated with "Cold Fusion...."

The following "Hugh Vreeland" article, under the same title, which is rated by Helium as a better article, was written later and added a little different angle.
  Let us try to extract it from the Helium site.
 Summarizing and organizing, " Hugh" helped a great deal in organizing the thinking.

Motion in a matrix as a new model of the physical universe

by **Hugh Vreeland**

Having read and reread a number of times the lead, and, at this point, only other article on this subject, (which I shall take the liberty of henceforth calling "Doc's article,") I am daring to try my own version which I might call, "A Universe of Basic Black Holes."

In another article I have defined what I consider the simplest black hole, an entity which, if the postulate be correct, has a vibrational and rotational period which corresponds to the speed of light and which would be one of those which makes up a matrix of particles which would be in contact with one another without there being a spacing.

With a vibrational and rotational velocity corresponding to the speed of light, this contact would transmit information at that speed. In Doc's article, he assumed a spacing, and, inadvertently, calculated the dimensions of a black hole having a mass of one gram.
There are, however, useful conclusions, or implied conclusions, in that article which I wish to accept as basic postulates:

!. There is a matrix.
2. Motions can be analyzed in two categories.
3. Equating the two energy expressions is a valid idea. (E=(mv^2)/2 & E=hu combined to :(mv^2)/2 = hu.)
4. Mass is a vector quantity which probably measures the component of angular momentum opposite to the vibratory or translatory motion of a particle.
5. Planck's Constant, 6.63 x 10^-27 erg-sec is a measure of a basic angular momentum.
6. Mathematical manipulations, including differentiation and integration, carried out on the energy expressions may describe qualities of a fundamental unit.
7. The momentum equation obtained by differentiating the combined energy equation, that is, mv=h, can be evaluated inserting the values for the speed of light, "c," and Planck's constant, "h," to give a value for the mass of a the fundamental unit, a Basic Black Hole. (This is approximately 2.21x10^-37 g.) Since this was originally written it was realized that this mass is the mass of a black hole having a radius of 1 cm. A better version of this statement is to directly equate Planck's constant to the definition of angular momentum. When this is done it is seen that 2.2l x 10^-37 g..cm. is the size of a "universal torque or unit of angular momentum.
8. The momentum equation can be integrated to form two energy equations wherein either mass or velocity is considered as a variable.
9. One of these equations may describe vibration the other,
rotation. The assumption being here that the equation considering velocity as the variable describes vibration for a stationary particle. The mass equations then is related to rotation. To these I add a few more postulates:
10. The momentum equation from #7 above describes not only the Basic Black Hole, but also in the form of mrv=h, all basic particles including the electron and proton.

11. Energy calculations for the Basic Black Hole can be used to estimate other Black Hole aggregates.
12. Integrations of the energy expressions may be used to estimate other characteristics of natural units.
Now, having said all of this, it seems time to look at the two, or maybe four, other fundamental "Black Holes" that are mentioned in "Doc's" paper, the electron, positron pair and the proton, anti-proton pair. He considered that these may be the fundamental vortex particles, of motions in the matrix.

I have a different idea, taking off on his idea of the proton being an "exploded" positron. What if electron-positron pair production represents. not an aggregation of our basic black holes but rather a splitting of one of our "Simplest Black Holes" into two components! This would occur with the two halves spinning off in opposite directions with vibrational energy converted to translational and rotational energy. This idea does not seem to have been mentioned by anyone else, but to make sense.
Can anyone pick up on this and show that the above speculation may fit because of the ideas mentioned above or because of information from other sources? How about showing that this idea can not possibly be valid?
Hey, other people, the more input that comes in on the basis of our reality the better chance we have of understanding what reality is .

Note: Since this article was first written, it has been shown that the angular momentum equation taken directly from the fact that Planck's constant has the dimensions of angular momentum can be evaluated at the speed of light, "c," to give an equation, m x r = h/c, which defines a "family" of entities having a symmetrical "central" entity having "m," and "r" having the absolute value of (h/c)^0.5. This central particle, which "Doc" dubbed the "Sin-

Vree" Particle, can be considered, if one wishes to, the parent entity of almost any particle with the possible exception of the neutrino, anti-neutrino and any "parent" formed by their combination. At this point, an anti-neutrino/neutrino "parent" seem a good candidate for the dots of the matrix. The theory of this, is evolving, and, perhaps, I should not modify this paper too much more. who knows, someday it may have historical significance

What is the matter? Here is a little article about "Anti-matter." I had to edit it a bit here, as, on looking closely, the version posted on Helium.com is somewhat confusing in its description of a "Type one oscillator." I tried to correct that here.

ON THE MATTER OF ANTI-MATTER, What is hidden where?

The writer, noting a "Hidden-half-of-existence" below a radius of 4.7×10^{-19} cm., and the possible creation of alternate universes resolves the 'lost anti-matter problem of our universe " to his own satisfaction.

Re: Matter and Anti-matter.

Every one has heard of Matter and Anti-matter. Matter is made up of electrons and protons and anti-matter is made up of anti-electrons and anti-protons. When anti-matter and matter come together to meet they mutually annihilate to form pure energy. That is the commonly accepted picture. The reality may be not quite that simple, and, possibly, more interesting.

There is a distinction in science between something which is dubbed "Matter" and "Anti-matter." Matter is made up--for the most part-- of electrons and protons while anti-matter would consist of the corresponding "anti-particles." There seems to be a problem, however, as to whether anyone has actually observed an anti-matter atom or molecule. The writer does not personally know of any reports of anti-Hydrogen or anti-Helium, etc. There is a rather extensive writeup available on the Internet of a theory called "Dominion Cosmology" which does a quite logical development of a Cosmology based on the premise that matter and anti-matter have a property of mutual repulsion. There is, also, an interesting 2006 news release that the "Beta sub 2" particle vibrates (oscillates) between matter and anti-matter.

Let us look at the whole problem of "The Lost Anti-Matter of Our Universe." from the viewpoint of the Oscillator/Substance Model.

[For those who are unfamiliar with the O/S model, the following short introduction is given. O/S Model postulates existence as within a substance at/near its triple point, which is made up of (or organized into) oscillators /oscillations defined by the family set, $\{m \times r = h/c = r \times m\}$. That is, they--the organizational units--have a constant torque of h/c, Planck's Constant divided by the speed of light, and since the absolute values of m and r are interchangeable, and at an average value would be equal to each other and to $(h/c)^{0.5}$. All of these oscillators can be considered to invert through values of about 4.7×10^{-19} g. at 4.7×10^{-19} cm.]

If the "O/S" Model be valid, the problem of Matter/Anti-matter, is tied closely to presence of oscillators as the organisational units of reality.

Oscillators/oscillations within a substance can be classified into three general categories. (At least, that is the way this writer classifies them.) All three categories will appear as disturbances of a spherical form, but their "motion senses" will be different.

The simplest category, which we shall call a Type 1 Oscillator would be a true sphere, inverting from an outer sphere through an inner sphere-- in our Model, the inner sphere would be as defined above,
i.e., $(h/c)^{0.5}$--from an outer limit of maximal size and minimal measurable "mass," to an inner limit, also a sphere of minimal size and maximum "mass."

[Size is easy to understand. Mass, however. does not ever seem to have been given any true definition. As used here it will mean the pressure/tension between an entity and the rest of the "Substance of Existence" as measured at a point on the boundary between the entity and the "Rest." The more the concentration of motion within a space, the larger will be the measured mass. For a Type I Oscillator, indeed for any of the oscillators, the smallest mass would be corresponding to the largest size, and these would be the values which will be found in our "Reality," if they can be measured. It appears that most Type I Oscillators may go totally undetected.]

A Type 2 Oscillator could be called a "Toroidal Pseudo-Sphere." The total space occupied would be spherical, but the sphere would have, at any given instant, an axis of rotation and an equator. This type would invert through a circle rather than a sphere, and can, with sufficient motion content added to it develop counter rotating halves and, eventually, spit into two Type 3
oscillators

'

Type 3 oscillators would be always be formed in pairs having opposite senses of rotation/pulsation (inversion). The best known of these sets are the negatron-positron pair, also known as the electron and anti-electron. Type three oscillators because of their effects in the Substance being opposites are known as "charged particles." Here is where the matter/ anti-matter concept arise. A unit of once rotation/inversion pattern is the negative charge, the reversed pattern is the positive charge. If the two patterns are of the same oscillation limits, they can rejoin with loss of energy in what is known as "annihilation." If, however, the oscillation limits are different, the two units may associate in may ways, but, do not rejoin to form a Class I or Class II oscillator/oscillation. In our universe. we have an observed unit, the neutron, which splits in to "halves" which are not identical, these halves, the proton and electron can associate to form may things, but do not rejoin to the neutron, nor can they rejoin to what one may call a "zerotron," a unit which is can be postulated as a Class I oscillator which can be split to an electron/anti-electron pair or deformed into a neutron which then splits into an electron and a proton.

The proton and electron are considered as matter particles as they have apparently indefinite lifetimes in our Universe. The reversed particles would be anti-matter, having indefinite lifetime in an alternate universe having some reversed sense. In O/S thinking, our universes has one rotation/pulsation orientation which is compatible with the rotation/pulsation orientations of the electron and proton. At the instant of inversion, or splitting, wherein our Universe emerged there would have been a complementary Anti-Universe also emerge. This can be used as one explanation for the lack of observation of "Anti-Matter" in our Universe.This, in the main is very possibly correct. However, there is another explanation. It may well be that the idea of Matter/Anti-matter is an over simplification, and that, in essence, the halves of nature exist in both universes.

 It, also, may be noted that the idea that matter and anti-matter will annihilate on contact may be in error. If matter and antimatter more complex than the simplest opposites, were to be formed in our Universe there would be slight differences due to orientation to the rotational characteristics of our universe, the exact "orientation fit" necessary for the "Yin-Yang" rejoining necessary for "annihilation" might be difficult. It is known that the positron-negatron pair exists for a period of time as a "Hydrogen type molecule" before combining with loss of energy.

 Incidentally, it may also be noted that the usual statement that there are "s two photons emitted at 180 degrees" seems a misapprenhension to this writer. A point radiator will radiate one pulse in a spherical pattern. In the case of the "annihilation" recombination, what would be emitted would be one pulse, "photon" at 360 degrees. It could be detected as "two photons at 180 degrees," but this would actually be only one "energy" pulse.

It makes sense in terms of our oscillators to consider that what we observe is the maximum size, minimum mass part of any given oscillator, the size being greater than 4.7×10^{-19} centimeters and the mass less than 4.7×10^{-19} grams. There being another "half" to the proton, or the electron, or any other oscillator which we can observe which is smaller the the values quoted above but grater in mass. That is , every component of our existence has a duality. The 'anti-" half of the electron "hides" at a size smaller than the "4.7 limit" as does the other half of the proton, which could be considered as a different form of an "anti-electron." [It can be shown, mathematically, that if an anti-electron were slowed to about 1/42 of its velocity in a given direction, with all of that linear kinetic energy converted to internal rotational-vibrational energy, i.e, "Mass," it would become a proton .

The conclusion of all of this is that the answer to where is the antimatter which was created here in our universe is, if you consider anti-matter as having been created in our universe, most of the anti-electrons didn't get created here, protons showed up instead, Also, since there is another half to everything, which we do not normally observe one could consider that the other half is hidden there.

As long as the oscillators in our Universe stay in their most usual patterns, there is little observation of what are called Matter-Anti-Matter type interactions. Only on unusual happenings do we get glimpsed of the "other side of existence." The processes of positron ejection in certain types of "nuclear decay" and 'annihilation" and "pair-production" offer glimpses of the hidden, unknown parts of existence. Consideration of that possible, hidden half of oscillators, and of the possible ubiquitous presence of the "zerotron" predecessor of both positrons and negatrons and neutrons. could lead to a whole new view of Reality.

Time Out...Here is a reprint of my article on "Time" which was published on Helium.com. Clearly, the Helium.com raters didn't like it very much.

Essays: The concept of time

by **Dean L. Sinclair**

Time is a human construct used to keep track of sequential motions in space referenced to repetitive observed motions. Useful for planning and information transfer, it is sometimes considered as if it were a fourth spatial dimension.

Time is usually considered to have three aspects, the past, present and future. These can be considered to correspond to motion of a specific point on a wave front. The past is the total sum of all motions that led to that instantaneous position which we consider the present, and the future is where that wavefront spot may be considered to go in the next instant and all the instants which may follow.

The present is the result of a certain sequence of motions, which we call the past, something which no longer exists, but which is never the less a "fixed construction." To go back into the past, in a physical sense, would require a retrograde repetition of all of these motions, a set of motions which would increase instant by instant in a huge geometric progression. Our "wavefront" would have to move backwards in a perfect reversal of the sequence by which it had previously moved forward. Even were this possible, one can see, that, since the direction has been reversed, what was "back" is now "forward," the wavefront, while "retracing the past" would actually by "moving into the future." In trying to go into the past, a "time-traveler" would actually be attempting to create a future which was a reversal of the past.

If one considers that it may be possible that long wave fluctuations in the "Matrix" in which we exist creates multiple adjacent universes in which certain sequences may coincide, it might be possible to move from one alternate universe into another which would correspond exactly to some point in ones own past. There is no real indication that there are such alternate universes; and, were they to exist, it is highly unlikely that they would correspond in such way as for there to be possible entry from one to another.

However, if one wished to combine Religion, Brane Theory, and ultra-slow vibrations in the Matrix of Existence, maybe "Heaven is Just a Brane Away?" That sounds like a new hymn, or a Country-Western Song.

Here is another effort published on Heliium.com which has fared a little better. Unlike "The Resident Lunatic," this effort was not submitted to Poetry.com

Poetry: Existence theories

by Dean L. Sinclair

Some limerick style thoughts, dedicated to scientific theorists.

It may be that on a far-gone day,
Someone set up a 3-D dot array,
Picked the right spot,
Flicked a dot--
And Creation was on its way!

The above may not be right;
But, isn't it a fright,
That tangled threads,
Can mix up our heads,
With no logical ends in sight?

Of both Motion-in-Matrix and Thread Theory,
Any sensible scientist should be leery.
If we could put both to the test,
Possibly one could work best,
Is all that I can see clearly.

Yes, the results can be delirious,
Although the subject may be serious,
When one tries to say,
In a light-hearted way,
Thoughts on a subject mysterious....

Here is an article written some time ago that I liked.

A different model for atomic orbitals   1 of 1

Dean L. Sinclair

While atomic orbitals are often understood in the Bohr model as if they were analogous to planetary orbitals, or are explained as wave motions in quantum mechanical terms, there is an unpublished approach to understanding electronic orbital structures which might be of interest to some chemistry teachers or theorists. This "3-D Pendulum Model" as a theoretical approach for electronic orbitals, is very easily understood and makes perfect sense.
Although it was used in his classes by a chemistry teacher in North Carolina in about 1961, it never seems to have gotten into the literature as any part of science theory.

The idea is this, an electron attracted to a proton, or a nucleus, would move toward it at an ever increasing velocity, much as a pendulum is attracted to the center of the Earth, but would pass right through the center of the proton or nucleus, changing direction slightly in the passage, as both the protons in the nucleus and the electron are spinning bodies, go out the other side slowing down to a "stop" then turn back to repeat the process, after a number of turns a pattern would be established of electron motions. Since the greatest amount of time of the electron would be spent very near where it "stops," a probability picture of where the electron could be found would be a "shell" or "orbit." This, of course, is exactly the same conclusion as is reached by other models of electronic structure. This seems to be quite strange if one considers the nucleus of an atom as a very hard ball and the electron as being something light and fluffy. If, however, one does the mathematics on the listed masses and volumes of electrons and protons, it can be shown that the electron actually has a far greater mass density per unit volume than the proton and could easily pass through protons, and hence atomic nuclei. Once this is understood the idea of the electron moving much like a "three dimensional pendulum is more easily understood than the more abstractly mathematical approach of quantum mechanics.

To understand the idea better, one can imagine some sort of alternate reality in which something dropped on the surface of the Earth could be attracted by gravity to the center of the Earth and pass right on through without being impeded by friction. Exactly the same type of motion would result. The object would pass through the Earth to a distance equivalent to its starting point, and then fall back though again, and again, forever.

The idea of a body moving in three-dimensions in this manner is most easily understood, in discussion, as above, where but two bodies are involved as would be the case with an object and our "alternate earth" or an electron and a proton in the Hydrogen atom, but can be extended to describe situations where in two objects will repel each other while being attracted to a third. This would be a simple model of a Hydride ion. It can be seen seen that the idea can be extended almost indefinitely and should be possibly useful in computer modeling.

A mathematical wizard student of the aforementioned chemistry teacher said that the mathematics of this 3-D pendulum model works out to be the same as the quantum mechanical model. This writer is not a good enough mathematician to verify this, and the present address of Douglas Rogers Sandy Dymphus Christoper Messer is unknown, if he still be alive.

Note added July 2009.

Later examination of data showed that the electron can be considered either as much bigger and lighter than the proton, or much smaller and denser, depending on which limit of its vibration one is considering... Apparently, which idea one gets depends upon which part of the literature one studies.

An interesting sidelight, Sandy saw the article on Helium .com and got into contact after 25 years.

Let's pick oup another topic which was published on SciScoop; however, we'll just pull it frm our own file

Commentary, Social Science and/or Economics

The view is taken that two interacting Laws of Nature would have to be very consciously considered by human leaders for human society to have an indefinite life span.

A poker tournament on a space ship is used as an analogy.

Civilizations of the past have reached great heights of sophistication and accomplishment, only to disappear. Present human civilization its rapidly accelerating in knowledge, accomplishment, numbers and interdependence. Considering the effects of two seemingly obvious, but, apparently, essentially overlooked, factors, it is, speeding more and more rapidly toward a disastrous collapse.

The first of these factors involves what scientists would call the "First and Second Laws of Thermodynamics." These are sometime described, respectively, as "You Can't Win;" and "You Can't Even Break Even." These statements taken together could be called "The Law of Finite Resources." "In any space there is a limited quantity of usable items. When anything is done with these some are converted to unusable forms." Most obviously, and, possibly most critically, this applies to Energy--the quantity of "motion packets" available to humans to carry out what they wish or need to do. It, of course, also. applies to other necessities, water, air, shelter....

The second factor, which also should seem obvious, but appears to be overlooked is the distribution/redistribution of wealth by the Laws of Chance. The Laws of Chance redistribute wealth--material objects or symbols representing them--toward those who already have the most wealth until an equilibrium is reached wherein wealth will go in and out of the "biggest pile" at the same rates. This equilibrium is, however, only possible if there is an external source of wealth to compensate for operation of the "Law of Finite Resources." Transfer of wealth involves a transportation factor which involves use of energy, hence the involvement of the Law of the Finite..."

In human terms, the old "Lord and Serfs" arrangement worked--sort of--as long as the serfs harvested enough energy from the Sun to compensate for what was needed to keep them alive and to maintain the Lord at his needed level. If the Lord got too greedy, the situation would rapidly deteriorate. With too much wealth going into his hands, not enough left to maintain the serfs, the arrangement would rapidly collapse.

Let us look at the redistribution of wealth bu the Laws of Chance from the viewpoint of a poker tournament. Envision a starting point of players of equal ability with equal stakes. The person who lucks into the first really big pot at any given table will almost surely break the table. The winners move on and the tournament continues until there is one winner with all of the stakes. The game is ended, unless the stakes are redistributed to start over.... [In real life, such a winner might well go out into the streets and get mugged.] Place this poker tournament on a Space Ship where both the stakes and the other necessities to keep the games going have to be found within that ship with no outside source of supply, in other words by dismantling and "eating" the ship, and one has a fair analogy to what is happening on our Earth.

If an outside source were available which could furnish the energy resources to keep the players alive and to replenish the wear and tear on the stakes, the tournament could go on indefinitely, as an Equilibrium Condition such as has been described before could presumably be attainable which would continue as long as the external source were available. If the input were actually more than needed additional players could be added or the stakes of the game raised.

Earth has such an External Source available, Sol, the Sun. Were we humans harvesting at least as much energy from the sun as we are using, we would stand a chance of maintaining our civilization, indeed, our very existence indefinitely. Possibly, even to the point of finding another source among the stars before Sol goes Nova. That is, of course, if we do not destroy other critical resources such as air, water, habitable land space....

The chances, however, of we humans being able to put together a recipe for indefinite survival are probably slim to none. Some of this is being written on the Day of Inauguration of Barack Obama as the 44th President of the United States of America. The faith of the World is at this moment resting on the shoulders of this brilliant man. One can only hope that somehow he can pull off a miracle of convincing the peoples of the world that we are in this together, that there is no space for fighting over ideologies when that fighting can only hasten the destruction of human civilization, hastening the end of the human race itself, and leaving nothing behind to worship and glorify the God or Gods, if there be such, that the fighting was in the name of. He and the other world leaders must realize that increased GNP merely means faster destruction of resources, that human conflict and wars are total wastes of resources. There is no place for human greed, if humans are to survive indefinitely....

The two "Laws" of which this writer speaks, have been too long ignored. It probably is far too late to keep our "game" going indefinitely, However, isn't there some chance that it could be prolonged a little if leaders could show a little awareness and cooperation?

This was a letter to a newspaper which got the expected response--none what-so-ever...

A Letter to Sophie..

To the city desk at the --------   --- -------------

This is not exacty what one would call a "news release." It is simply some thing which I hope you can get enough of a "kick' out ouf to pass it on to the people who do the Judge Parker Strip.   Thanks.

(Of course, if you thought it was fun enough to publsh otherwise. you have my permission....)

To Sophie Spenser c/o Judge Parker, c/o Seattle Post-Intelligencer

Dear Sophie.

As my favourite heroine, I cannot resist trying to get you involved in some of my world. I think I owe it to you and to all the brilliant youngsters like you, to give you a chance to look at some ideas and see what you could add or change. I'm sure that with your razor sharp intelligence, you will understand immediately what it took this writer over five years to figure out, or 77 years if one wanted to look at it in another way....

Here's the situation. If one switches/reverses the interpretation of two sets of over a Century old data, one can put together the basis of a quite comprehensive "Theory of Existence, " which seems to be able to be extended to account for almost everything, except, of course, the fact of "Existence."

 What is done is to reverse the interpretation of the famous Michelson-Morley Experiment which determined the Speed of Light to be an apparent constant of nature, from proving that an "Aether, " an all-pervasive something, does not exist, to the idea that it actually furnished information about that "Aether."   The other switch is in the naming of Planck's Constant which connects the frequency of light to Energy, from a "Constant of Action" to the mathematically equivalent, "Constant of Angular Momentum."

The Speed of Light, as determined in the "M/M' Experiment, can be shown to be the maximum velocity of the carrier wave information in a medium. That it was the maximum velocity for information has always been known, but, it's significance has been ignored. The Maximum Velocity at which information can be carried from a point is the Average Speed of the Carriers in any given direction.   The Speed of Light,then is not necessarily a limit, it is certainly an average. That is.if movement of information is the same in principle by electromagnetic radiation as by Pony Express. The "M/M" Experiment can be taken therefore to have determined a characteristic of the Aether, as being able to carry information by light waves as the Speed of Light, "c," which is about 3×10^{10} cm./sec.
Also, it showed that the "All-pervasive-substance," could act as a solid, in order to carry transverse waves. A moments consideration of a fact of chemistry leads to the conclusion that a substance at its triple point where it can be solid, liquid or gas on slight changes of pressure fits this very well. Also, as a substance at its triple point would have the nearest situation to a total equilibration of motions throughout, and would tend to revert to that

situation, a solid at its triple point accounts for such things as the Law of Forces. For each and every force there is an equal and opposite force." One can paraphrase this, " In a substance at its triple point any disturbance will be compensated...."

There would always be motion in a substance. Planck's Constant as an Angular Momentum implies that an important part of that motion would be rotation. as Angular Momentum is a characteristic of a rotating unit. If we set Planck's Constant, "h,h" equivalent to the definition of Angular Momentum, Mass times radius times velocity, we can determine some characteristics of at least some of the substance. We write, $h = m \times r \times v$. Evaluating this at "c," and rearranging we get $m \times r = h/c$. Since mass times radius equals torque which is the push or pull on a rotating object, this makes sense. We have discovered the "torque constant of nature." Additionally since the equation can be fitted into the form , $xy=K=yx$, which is. among other things, the definition for an oscillator, we have discovered an equation defining an oscillator set, which we can write in set notation, {$m \times r = h/c = r \times m$) , which points out that the numerical values of mass and radius can be interchanged and at some point they will be equal at $m = r = (h/c)^{0.5}$. This last figure is about 4.7×10^{-19} cm. and 4.7×10^{-19} grams. We now have our substance made up, at least in part, of oscillators which invert through a constant value. We even have discovered an interior "dimension" to these oscillators that is smaller Than 10^{-18} cm. in diameter. This figure is the size of the hole into which string theorists say that their dimensions disappear and Quantum Mechanics is said to fail.

Checking published data for the electron and proton which are the basic units of all matter, we find that the Rest Mass and "Compton Wavelength" values fit as one set of limits, at the situation o maximum size and minimum mass, the switched Absolute Values define the balancing limit of maximum mass and minimum size....

We can go on from here with the implications of a substance at its triple point made up of oscillators of this family to develop an entire Theory of Existence. I'm sure, Sophie, that with your intelligence, in your world, you could get someone to listen to you and add, amplify, maybe even refute. If you can find use for any of the work this writer has done in your world, or you can find some way to help, you can have some of your contacts in my existence check http:// groups.google.com/group/oscillatorsubstance-theory.

In my world there is too much information, too much "white noise" for anyone to hear anything. Scientists are too busy creating new data to re-examine old data for new significances....

Sophie, if any of this could help you to perhaps get a jump-start on a career as a scientist or a full-ride scholarship in college in your world, you're welcome. In my world, if noticed at all. it will probably be ignored as "Cracked Pottery."

Cheers,

Dean L. Sinclair
P.O.Box 592
Aberdeen, SD57402-0592

Note: To Editors, ------- ------------------,

After working on a set of ideas, I realized that a bright, high-school student who was creative enough to not be "brain-washed" in to accepting, "It's in the book, and has been for a century, so it 'True.' " could easily come up with the same ideas. Sophie, if she were in our "Universe," would be such a person as to be likely to do such a thing....

I couldn't resist this communication....DLS

SWILL LAVETO RY TO FID IT. OMEHOW THE REST OF THE LETTER HAS SOMEHOW
Who knows? This article maight have been read bh somebody.....

The Hadron Collider and O/S Theory

(This first part was written about July 2008.)

The hadron Collider may have some useful results for O/S theory. If it doesn't cause a disaster; which O/S theory suggests is a definite possibility, some of the other result may fit in. It may be projected that the collision expected of two proton beams may not result in the release of the energy content in some explosive manner but merely in the divergence of the beams because of the spin effects on vortex particles. However, if there is energy transfer on such contact there is certainly enough content to cause electron-proton production, (some 1800 pairs or so) or even, possibly, to form one neutron. We're talking about the energy content of two protons accelerated to essentially "c" of course

If collision is elastic, with energy transfer, there will be an attempted acceleration of some of the protons to beyond the speed of light. This would probably have the effect of increasing their mass, which could result in the formation of very short-lived "bosons" which would decompose....

This writer still hopes that the entire system does not go into resonance with disastrous results.....

In any case, I still feel that there is a good chance of having "unexplained" Hydrogen atoms appear in the system....

Comment, July 23, '09. Approximately a year later,

Additional information has arisen making the above analysis seem somewhat naive. The Hadron Collider burnt out a component shortly after start up, apparently in early attempts at passing a proton beam around (more than once?) Anyway, it apparently hasn't started up again.

Additional ideas has developed in considering the proton. It was noted above that the vortex nature of the proton would cause problems. A little more consideration, taking in the fact that protons can interact by pairing processes and even further linkage, just as is possible with electron, shows that most likely a stream of strongly accelerated protons would develop either a linked structure for the entire chain or separate groupings corresponding to the "nuclei' of simple atoms. In either case, there is a definite possibility that the effect would be of a rotating, organized charge. This may well be the cause of the break down.

Additionally, the clockwise rotation of the proton would make the energetics of spinning the protons about a circular accelerator in a clockwise direction one situation, while spinning them in the opposite direction would be a totally different kind of energetics. There would be several areas of conflict, even with the ccw. spin of the Earth, and even the (Probably also ccw) spin of our Universe.) The approximation of the proton as a "charged point particle" is grossly misleading . If the insights of the O/S model are valid, it is very doubtful if the Hadron Collider will be able to go into operation, as the scientists and engineers have no idea of the problems they may actually be facing.

Comment 3unew 9,2013. After several restarts and, perhaps, some definte retooling of their instrument, the people at CERN announced that the had found the Higgs boson and went onto accellering Lead ion, at the last reportthat I have seen.. I doubt that they found the Higgs, but thefound something.... Don't kknow if they took my commets to heart or not....

Here is a "Rant" that was published in 2010. Situation hasn't changed much in 3 years.

WHOOPS! THIS IS THE END OF VOLUME ONE. THE COMPUTER IS HAVING TROUBLE WITH THE MANUSCRIPT BEING TOO LONG...GO TO VOLUME TWO FOR THE RANT AND ARTICLES OTHER TOPUCS IN ADDITION TO SCIENCE AND MATH...dls

ihis,is, of course, when, and if, the second half   (Volune Two) ever gets put together.....

www.ingramcontent.com/pod-product-compliance
Lightning Source LLC
Chambersburg PA
CBHW080552190526
45169CB00007B/2748